THE
EVOLUTION
OF AN
EVOLUTIONIST

THE
EVOLUTION
OF AN
EVOLUTIONIST

C.H.WADDINGTON F.R.S.

CORNELL UNIVERSITY PRESS
ITHACA, NEW YORK

First published 1975 by Cornell University Press.

This edition is not for sale in the United Kingdom.

International Standard Book Number 0–8014–0882–2

Library of Congress Catalog Card Number 74-10415

Printed in Great Britain by
W & J Mackay Limited, Chatham

PREFACE

The title of the book can be interpreted in two ways, and both of them are relevant to its content. This is, firstly, an account of the way in which the theory of evolution is seen through the eyes of one person who has been studying and working on it through the last three decades or more; it also provides the bare bones of the story of how he came in the first place to view it in this still somewhat unconventional manner, and how he gradually developed a greater confidence in the importance of this particular theoretical framework. It is, if you like, an exposition with some autobiographical background.

Three basic changes in 'paradigm' (in Kuhn's sense) of the Theory of Evolution have become accepted since the rediscovery of Mendelism at the beginning of this century; a fourth change is still waiting in the wings for full acceptance.

The paradigm shifts which have become orthodoxies may be summarized as:

1. Variation between individual organisms is due to changes in discrete units (genes) which do not 'blend'. (Paradigm; *individual* genes in *individual* organisms. Bateson, Morgan.)

2. Evolution is to be considered in terms of changes in frequencies of *individual* genes in *populations* of organisms. (Haldane, Fisher, 'Neo-Darwinism'.)

3. Evolution is concerned with *populations* of genes (gene pools) in *populations* of organisms. (Sewall Wright, Dobzhansky.)

In all these paradigms, only lip-service, if that, is paid to the distinction between genotypes and phenotypes. Natural selection is treated as though it directly affects genes and their frequencies, while the gene frequencies merely react to, but do not influence, the selective pressures.

Since about the time, in the late thirties, that paradigm 2 was becoming superseded by paradigm 3, I have found myself impelled to envisage evolution as dependent on processes which affect phenotypes

and have only secondary repercussions on frequencies of genotypes. This is perhaps a more profound change of paradigm than
the others, since it demands radical alterations in some of the most
deeply ingrained biological dogmas:
1. It points out that, although an 'acquired character' developed by
an *individual* is not inherited by its individual offspring, a character
acquired by a *population* subject to selection will tend to be inherited
by the offspring population, if it is useful.
2. It argues that genotypes, which influence behaviour, thus have an
effect on the nature of the selective pressures on the phenotypes to
which they give rise.
3. It introduces into the theory an inescapable indeterminism quite
different in nature from quantum intermediacy, but almost as
intractable, since identical phenotypes may have different genotypes, and identical genotypes may give rise to different phenotypes.

All these factors have important repercussions on the general
'philosophy' of evolution. This may perhaps be summed up in the
statement that, whereas the import of the previous evolutionary
theories can be sloganized in Jacques Monod's phrase 'Chance and
Necessity', the fourth paradigm would substitute slogans such as
'Learning and Innovation', or 'Adapting and Improvising', or, if
you like a more with-it jargon, 'Recompiling and Heuristic Search'.

The Neo-Darwinian paradigm, of selection acting on genes, is
good enough when we are considering situations in which fitness is
related rather directly to the qualitative nature of primary gene
products. Recent research techniques have made several categories
of phenomena of this kind amenable to study; for instance, enzymes,
immunoglobulins, tissue antigens, respiratory pigments, and so on.
The widespread occurrence of population polymorphisms for
primary gene products of these and similar kinds has certainly
presented a number of very challenging evolutionary problems; and
in meeting these, it is probably justifiable to follow the Neo-Darwinist
prescription, and leave out of account any effect of the environment
in modifying the phenotype on which selection acts.

Where this prescription fails to pass muster is just in problems of
evolutionary adaptation at the organ or even tissue level; and these
are the 'classical' problems of evolution. Moreover, they are the
problems which have philosophical implications. It is doubtful if
anyone would have ever felt any need to resist the notion of evolution
if all it implied was that the exact chemical constitution of haemoglobin gradually changed over the ages. The importance of the
Theory of Evolution, as part of man's thinking about the universe

in which he finds himself, is that it offered an alternative to the notion of an Intelligent Designer who had designed fish adapted to swimming fast (mackerel) or to lying flat on the bottom (plaice), birds adapted to flying in a number of different ways (hawks, sparrows, humming birds) or to not flying at all (ostriches), and animals adapted to live out their lives in an almost infinite variety of different ways, but always very effectively.

It is where it comes into conflict with the Theory of Intelligent Design that the Theory of Evolution becomes something of general human importance, rather than a mere piece of technical specialized expertise; and it is just in these areas of conflict that the Neo-Darwinist paradigm is misleading. The successes of Neo-Darwinism, pepped up as necessary with a shot of stochasticism, in explaining the evolution of the haemoglobins, nuclear histones, or histo-compatibility loci—for which we are still waiting, but with good grounds for expecting a favourable outcome in the not too distant future—will be largely beside the point when we are considering Evolution as a relatively new, major component of man's thinking about his place in the universe.

This book is built up from a selection of articles which I have already published. They have appeared in a wide diversity of technical journals and other perhaps more obscure places, such as proceedings of various symposia and conferences. They have in a few cases been lightly edited, to cut out some of the more boring details of the stocks used for the experiments, the technical methods employed etc., but they are reprinted fully enough to exhibit the degree of validity of the conclusions drawn. The rest of the preface shows how each article, or group of articles, takes its position within the developing body of theory and experiment, which is the essential subject of the book. To my eye at least—which is the eye of someone who is basically a professional embryologist, who sees every present item as one frame in a continuous movie—they cohere together into a rather definite unity, though, of course one with rather fuzzy edges. In fact, looking back over a lifetime's interest in evolution, one might be a bit depressed by the extent to which this unity justifies T. S. Eliot's dictum: 'In my end is my beginning'—with its implication that there is not much more than the beginning. Did I really have all my main ideas before I was forty? Perhaps, but after all, I got all my genes when I was aged about -0.75 years. The genes have not finished doing their stuff yet; and the ideas I think have also been slowly developing and working themselves out more fully. At least, I like so to kid myself.

The first paper, although written quite recently, is a reminiscence about my first studies in evolution in the late 1920s, when, as a palaeontologist, I collected and classified some faunas of fossil ammonites. It gives something of the background of interest in the philosophical ideas of Whitehead which encouraged me to take up such a subject in the first place, and in particular why I chose to study a group of animals in which the developmental history of individuals is open to inspection. But, to my mind then, it was not inspectable in sufficient depth, and for about a dozen years I left evolutionary problems on one side, to engage myself with the causal analysis of development and its connections with genetics.

Papers 2 and 3 record a renewal of interest in evolution at the beginning of the forties. I was looking at evolution very definitely from the point of view of a developmental biologist. In paper 3 I suggested that there should be a mechanism by which selection could convert what had initially been a conventional 'acquired character' into one not dependent for its appearance on any particular stimulus from the environment. This mechanism, which I later christened 'genetic assimilation', was at that time purely theoretical.

The next group begins with the first announcement that genetic assimilation can occur in practice, just as theory had predicted (4). It would be tedious to reprint here all the experimental work which followed up this initial lead. Paper 5 is an early discussion of its implications for evolution in general; 6 is an interlude, a brief reference to one of the philosophical consequences; and 7 gives a more developed view of its biological role, written some seven years later, and also provides illustrations of the experimental systems which I used. A comprehensive account of the whole series of experiments in genetic assimilation is given in 8, where there is a full bibliography for those professionally interested; while 9 is a description of the interesting observations and experiments on a case of genetic assimilation in the field made by the great Swiss psychologist Jean Piaget, which are not as well known in the English-speaking world as they deserve to be.

Item 10 is a correspondence (which I hope rejoices the heart of Professor Robert Merton, who seems to think that questions of 'priority' are the most important aspect of the sociology of science) about the work of the Russian biologist Schmalhausen, who was thinking along lines very similar to those I was following, at about the same time and quite independently.

Papers 11, 12, and 13 deal with one of the factors which makes possible the process of genetic assimilation, namely the canalization of developmental pathways. This is followed by a group (14, 15, 16) of papers which have tried to give more solid factual support to the idea that the genotype may have an effect on the character of natural selection by influencing phenotypic behaviour, both in respect to the physical environment and in such inter-individual relations as the choice of mate, while 17 is an exchange of letters with Professor Sheppard, the main point of which is again the importance of behaviour in modelling the character of natural selection.

The surprising reluctance of orthodox evolutionists to realize this also emerges very clearly from the first item of the small collection of reviews brought together in 18. Two minor, but perhaps not uninteresting points follow as 19 and 20.

An attempt to formulate the 'post-Neo-Darwinist' view follows in the next group. Paper 21 is concerned with the special case of 'colonizing species', that is, species which have recently moved, or been moved, into regions where they were not previously present, or into habitats which have arisen only recently, usually through man's activities. A general principle which has been somewhat underrated in the past, is stated in 22. 23 provides a very elementary mathematical model of some of the basic ideas. Then 24 and 25 are extended statements of how I view the theory of evolution today.

Paper 26 is a further attempt to consider how our ideas about evolutionary processes need to be modified to take account of the facts of developmental homeorhesis.

Finally, there are three papers (27, 28, and 29) which discuss the very special character of the evolutionary processes which are important in the history—past and future—of mankind as a culture-developing animal.

Many of the topics discussed in this book have been dealt with also in earlier books, particularly *The Strategy of the Genes* (1957) and *The Ethical Animal* (1960). The material in this volume both brings these ideas up-to-date, and also, by including writings from a period even earlier than those books, shows how they originated and developed.

The papers included here are by no means everything that I have written about evolution and topics related to it. Nearly all the material in this book has been published before; the exceptions are papers 9, 10, 17, 18c and 29. I am very grateful to the Editors of the publications mentioned in the Acknowledgements for permission to reprint.

C. H. WADDINGTON

Contents

1

THE PRACTICAL CONSEQUENCES OF
METAPHYSICAL BELIEFS ON A BIOLOGIST'S WORK.
AN AUTOBIOGRAPHICAL NOTE

Several of the more 'hard-headed' characters at the Second Symposium [1] expressed from time to time, at cocktails or after dinner, a suspicion that metaphysical considerations of the kind introduced by David Bohm have ultimately no real impact on the directions in which science advances. They suggested that they were merely part of the froth churned up while the theoretical physicists flounder and thrash about trying to find a firm footing in the deep and dangerous waters of quantum theory, sub-nuclear particles, and the like; and that when this footing has been found the froth will settle down and disappear. It is software, they suggested, so soft as to be deliquescent and ultimately evanescent.

I do not agree with this. I should like to argue that a scientist's metaphysical beliefs are not mere epiphenomena, but have a definite and ascertainable influence on the work he produces, by reminiscing for a moment about my own career. I am quite sure that many of the two hundred or so experimental papers I produced have been definitely affected by consciously held metaphysical beliefs, both in the types of problems I set myself and the manner in which I tried to solve them. I do not want to argue now—though I'll do so later if anybody wants me to—that these were really the most interesting problems, and that I set about them in the right way. Maybe my metaphysics was leading me up the garden path (though I don't think so); but the point I want to make now is that it was leading me somewhere and was therefore something more than a set of decorative flourishes on the proscenium arch, giving on to the stage in which the real action takes place.

As David Bohm points out, metaphysics is a sort of poetry; and we are all talking poetry all the time—if you doubt it, look again at what Richard Gregory has to say about such a basic activity as perception. As poetry, metaphysics can be absorbed through communication-channels other than extended rational exposition. So, to begin this metaphysical-experimentalist's autobiography, I will mention two (or perhaps three) notions which infiltrated into my thinking at a very early stage, without much benefit of academic dignity, and which have remained there ever since.

When I was a schoolboy there was a peculiar period after you had taken your entrance or scholarship examinations to the university in December of year n, and before you actually went to the university in October of year $n + 1$. This was the time when your schoolmasters could really have fun with you. I had been in the classical sixth form, learning mainly Latin and Greek and a certain amount of mathematics; but when it dawned on me that all my friends were leaving and I had better leave too and go to the university, I decided I had better try to get in on the basis of chemistry. Fortunately we had a chemistry master, E.J. Holmyard, who was something of a genius of a teacher. During one summer holiday and autumn term, he taught me the whole of chemistry, at least enough to push me into the lowest grade of assisted entry to the university, an Exhibition. After that he could really break loose and teach me what he was interested in. His passionate interest happened to be the functions of the Alexandrian Gnostics, and the Arabic Alchemists derived from them, in transmitting both the philosophical ideas and the technical knowhow from the Greek civilization which expired around AD 200, to the European one which began to come alive at the Quattrocento. So he made me learn a smattering of Arabic and look at a large number of very odd late Hellenistic documents. Two ideas stuck:

The world egg. 'Things' are essentially eggs—pregnant with God-knows-what. You look at them and they appear simple enough, with a bland definite shape, rather impenetrable. You glance away for a bit and when you look back what you find is that they have turned into a fluffy yellow chick, actively running about and all set to get imprinted on you if you will give it half a chance. Unsettling, even perhaps a bit sinister. But one strand of Gnostic thought asserted that *everything* is like that.

The Ouroboros, the snake eating its tail. This famous symbol, which was as well known in ancient China as in Alexandria, expressed the whole gist of feedback control almost two millenia before Norbert Wiener started 'creating' about the subject at MIT and invented the term 'cybernetics'. Here is a drawing of an ouroboros which I made for an essay which I wrote while I was still at school, presumably around 1923. I reproduce it because you will see that inscribed within the ouroboros is a third subsidiary notion; the slogan ἐν το παν', 'hen to pan', 'the one, the all', a phrase which implies (in a cybernetic context, be it remembered) that any one entity incorporates into itself in some sense all other entities in the universe (figure 1).

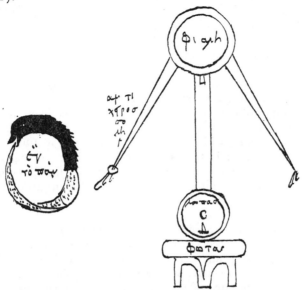

Figure 1. The Ouroboros, together with an alembic (distillation vessel) redrawn from an alchemical document known as the Chrysopeus (i.e. gold-maker) of Cleopatra.

Before these highly poetic metaphysics had any practical influence on my scientific work, there was added to them a large body of much more explicitly rationalized thinking; in the first place that of Whitehead, to whose writings I paid much more attention during the last two years of my undergraduate career than I did to the textbooks in the subjects on which I was going to take my exams. Later this was joined by some infusions of thought which claimed to be materialist—either 'fancy' (dialectical), which preceded Whitehead and seemed to me to be in the main left behind by him; or 'crude', the prime example being Morgan and his school, who insisted that the gene is not just a logical construct from Mendelian ratios [cf. Woodger's definition

$$\text{mend} =_{\text{Df}} \text{Aeq}'\hat{x}\hat{y} \ (x, y\epsilon \text{ whz. } \text{Apr}'\text{Zyg}'x \uparrow K \upharpoonright \text{Apr}'\text{Zyg}'y\epsilon 1 \rightarrow 1)],$$

but is just a simple lump of stuff. But one was anyway surrounded by materialists, and the whole of science was dominated by essentially

Newtonian conceptions of billiard-ball atoms existing at durationless instants in an otherwise empty three-dimensional space. It was, for me, Whitehead who suggested new lines of thought.

What was this Whiteheadian metaphysics? I will sketch very briefly what were the salient features in my eyes.

1. The raw materials from which we start to do science—or with which we finish the scientific testing of a theory—are 'occasions of experience'.

2. An occasion of experience has a duration in time (cf. David Bohm, 'there are no things, only processes').

3. An occasion of experience is essentially a unity. Any attempt to analyse it into component parts 'injures' it in some way. Yet we cannot do anything with it unless we do analyse it. Our first step towards an analysis is to dissect the unity into an experiencing subject and an experienced object. The dividing line between these two is both arbitrary and artificial. It can be drawn through various positions, and wherever it is drawn it is never anything more than a convenience.

4. The content of any occasion of experience is essentially infinite and undenumerable. Moreover, wherever we draw the line between experiencing subject and experienced object, the latter will always remain undenumerable. If this were not so, we should merely have to describe the totality of the content of the experience, and the experience itself would be created.

5. The experience, which Whitehead refers to as an 'event', has, however, some definite characteristics. Whitehead refers to these as 'objects'—the word which is usually used, and has indeed been used above, in a quite different sense. Definiteness of the Whiteheadian objects in an event implies that, although the event has some relation to everything else past or present in the universe, these relations are brought together and tied up with one another in some particular and specific way characteristic of that event (Whitehead was writing *before* quantum mechanics became a dominant influence in our thought, but compare these notions with the idea that a particle must also be thought of as a wave function extending throughout the whole of space time.) For this tying-together of universal references into knots with individual character, Whitehead used various different phrases at different periods in the development of his thought. For present purposes I am content to stay with the least far-reaching of these, when he spoke of the coming together of the constituent factors in an event as a 'concrescence'. Later he described the way in which an event here and now incorporates into itself

some reference to everything else in the universe as a 'prehension' of these relations by the event in accordance with its own 'subjective feeling'. This is a metaphysics very close to that advocated by David Bohm when he speaks of creativity, and argues that nature is more like an artist than an engineer. Privately my own thought runs along similar lines; and I think they may be extremely important to the way in which one behaves in one's whole personal life; but I do not see that they have had any direct influence on the way in which I have conducted experimental work, which is the subject which we are discussing here. As far as scientific practice is concerned, the lessons to be learned from Whitehead were not so much derived from his discussions of experiences, but rather from his replacement of 'things' by processes which have an individual character which depends on the 'concrescence' into a unity of very many relations with other processes.

So, without going into further metaphysical sophistication, let's get down to brass tacks. What did I actually do as a practising biologist, and how was this influenced by this metaphysical background?

I began to work as a palaeontologist, studying the evolution of certain groups of fossils. And I chose, as my main interest, a group which forces on one's attention the Whiteheadian point that the organisms undergoing the process of evolution are themselves processes. The Ammonites were cephalopods, related to squids and the Nautilus, which laid down spiral shells. The animal occupied only the latest-formed part of the shell and from time to time moved forward a little, leaving behind it the part it had previously inhabited. Thus following the whorls of the spiral shell outwards, from the centre to the periphery, one has a record of the whole life-history of the animal; it never appears just as an adult whose juvenile stages have vanished. The whole developmental process is preserved so that one cannot avoid examining it. And the process is, of course, complex, with many facets. On the surface of the shell there are 'ornaments'—ribs, knobs, tubercles, etc.—which change as development proceeds; the cross-sectional shape of the tube may change, and so may the closeness with which it is coiled; and behind each living-chamber a partition is laid down which meets the outer shell in a complex 'suture' which is one of the most characteristic features of a species (figure 2).

In most types of animal, in which the adult form of the individual replaces the younger stages, the only way to study the developmental history is to collect a large population containing juvenile as well as

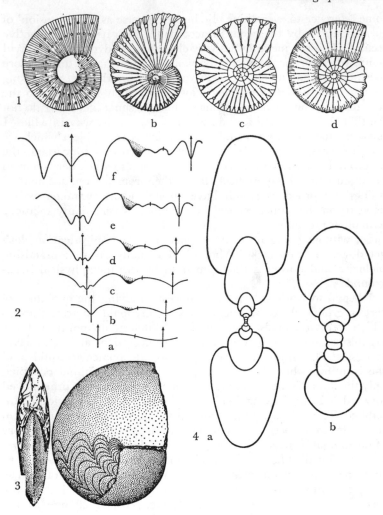

Figure 2. Ammonites. In 1 (above) a, b, c and d are an evolutionary sequence of species, to illustrate changes in the form of the ribs and knobs. 2 shows the development of the suture in an early species, from the young form (a) to the adult (f). 3 illustrates a very tightly wound form, and shows the sutures. 4 is a cross-section of a loosely coiled species, to illustrate the change in the shape of the spiral tube, the younger stages being shown enlarged at (b). (From L. *Mollusca*, in *Treatise on Invertebrate Palaeontology* (ed. Moore) vol. 4. Geol. Soc. America and Univ. Kansas Press.)

adult stages. Age can usually not be determined directly, but one can take some measure of size as an indication of it. Figure 3 shows a graph I made, around 1927, of the variation in ratio of breadth to length in a collection of fossil shells (a Brachiopod, *Terebratula*) [2].

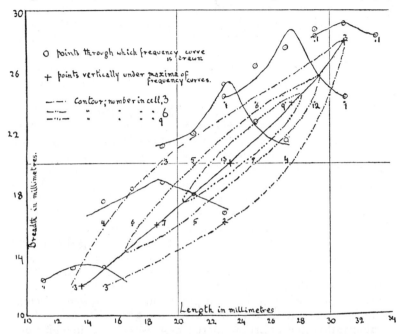

Figure 3. Frequency distribution of length and breadth in a population of fossil *Terebratula* [2], shown as a contour map with superposed frequency curves at five intervals. (From an unpublished exercise of my own, used in 1927 by my friend Reuben Heffer to try his hand as a printer in preparation for joining the family bookshop.)

These early exercises left me with a deeply ingrained conviction that the evolution of organisms must really be regarded as the evolution of developmental systems—which was the title of one of the first articles I wrote about evolution when I returned to the subject some years later. It is of course related to such old ideas as Haeckel's 'biogenetic law'—phylogeny repeats ontogeny—but I took it also as a guiding principle in population genetics. I still think that when modern population geneticists express the variation in a population by means of a timeless frequency curve, which deals only with the adults, this is a simplification which needs justification—

which of course it may often at least partially possess, for instance when one is dealing with the imagos of an insect like *Drosophila*, which metamorphoses suddenly into an adult form which thereafter does not change. But when I started doing experiments on *Drosophila* evolution, in the forties and fifties, I treated even that insect as a developmental system, and by manipulating the environment in which it developed was able to uncover the rather novel process of genetic assimilation. Thus my particular slant on evolution—a most unfashionable emphasis on the importance of the developing phenotype—is a fairly direct derivative from Whiteheadian-type metaphysics.

In my early career there was a considerable period in which I was concerned directly with developmental systems themselves, rather than with their evolution. My approach to experimental epigenetics was again strongly influenced by Whiteheadian metaphysics, but also by genetics. In fact, when I decided that I wanted to do something more experimental than is possible in palaeontology, I first tried to become a geneticist. My first two published papers were, one in plant genetics, and another a collaboration with J.B.S.Haldane along classical neo-Darwinist lines, in which we studied the effect of inbreeding on the segregation of linked genes, writing down, and trying rather unsuccessfully to deal with, a great series—up to 27 members, I think—of finite difference equations. So I had my taste of the thin gruel of mathematical formalism, as well as of the strong nourishing soup of time-extended populations of fossilized developmental systems. But my attempt to become a geneticist was a failure, because at that time in Britain there simply was no way in which one could earn a living at the subject. So I became an experimental embryologist. For some years my most immediate interest was in solving the technical problems of carrying out meaningful experiments on types of embryo which no one else had tackled successfully, such as birds and mammals, or of doing biochemical work on the very small tissue-fragments isolated from various regions of an egg.

However, the theoretical structure of the subject also needed a good deal of attention, and this is where the Whiteheadian approach came in. In the twenties, T.H.Morgan had argued that epigenesis should be considered in terms of the activities of genes, but he had little effect on the workers in this field. Embryological theories involved such notions as the 'segregation' or 'differential dichotomy' of 'potencies' to develop into organs such as the nervous system, kidneys, gut, etc.; or, at a more advanced but still very imprecise level, notions such as 'the organizer', 'induction', and the like. I wanted to return to Morgan's idea that the only 'potencies' it is meaningful to talk about are the potential activities of genes. So did several people who were primarily geneticists, but who had become interested in development without having actually worked on it very much.

Two rather radically different lines of approach were followed. The one which was—and in fact still is—favoured by most geneticists depended on what I think may be called an 'atomistic' metaphysics. It set out from the assumption of the existence of single genes, and it asked, at first, what does A do, and, later, what controls whether gene A is active or not? There is only space to mention, just to remind you, some of the key figures. Goldschmidt concluded that genes control rates of processes; Muller that they manufacture more, or less, or none of some substance, or sometimes an anti-substance, or even a quite new substance. Garrod (on human metabolism), followed by Haldane (on flower pigments), led on to the identification by Beadle and Ephrussi, in the early 1940s, of these substances as enzymes. Again in the forties, Hadorn in *Drosophila* and Grüneberg in mice studied the manifold development consequences which may follow an alteration in a single gene. Finally, about 1960, Jacob and Monod produced their story of the mechanism which controls the activity of single genes (or small operon groups) in prokaryotic organisms in which the chromosome normally lacks protein.

Clearly, this line of approach has paid off very well. But to my mind it does not really deal with the questions which the epigeneticist faces. For one thing, there is the difficulty that the Jacob-Monod control system could scarcely work, without considerable modification, in cells with proteinaceous chromosomes. But the problem is deeper. In cells of higher organisms we are not usually, if ever, confronted by the switching on or off of single genes. What we find is a whole complex cell becoming either a nerve or a kidney or a muscle cell. In the late thirties I began developing the Whiteheadian notion that the process of becoming (say) a nerve cell should be regarded as the result of the activities of large numbers of genes, which interact together to form a unified 'concrescence'. This line of thought had several ramifications. For instance, just before the Beadle-Ephrussi era I showed in detail how the development of the wing of *Drosophila* is affected by the activities of some forty different genes. Again a few years earlier it had become apparent that the 'gene-concrescence' itself undergoes processes of change; at one embryonic period a given concrescence is in a phase of 'competence' and may be switched into one or other of a small number of alternative pathways of further change—but the competence later disappears and if you've missed the bus the switch won't work. My main preoccupation, however, was with the nature of the switches, which was the subject of the experimental biochemical work I was doing, with Needham and others, on 'embryonic induction'. Influenced—probably over-influenced—by genetics I insisted that the switch must have sufficient specificity to recognize particular genes. We showed that, in these terms, the specificity resides inside the cells which react to induction—we called it 'the masked evocator'. This is very similar to the situation discovered by Jacob and Monod many years later in bacteria, where again the specific repressor molecules are internal to the cells which react to enzyme-inducing substances. If I had been more consistently Whiteheadian, I would probably have realized that the 'specificity' involved does not need to lie in the switch at all, but may be a property of the 'concrescence' and the ways in which it can change. Because of course what I have been calling by the Whiteheadian term 'concrescence' is what I have later called a *chreod*, a notion which Rene Thom has explicated; and the switches are Thom's *catastrophes*. The specificity *need* not be in what precipitates the catastrophe, but could reside only in the possible stable regimes (limit cycles in the simplest case) into which the system could be flipped.

Let us return to the beginning. The whole of this, I am afraid rather long-winded, exposition has not been aimed at showing that my line of thought, which I have derived from Whitehead, is the correct or best one. I should in fact admit that it has not so obviously paid off as the 'what does a single gene do?' line; although I also continue to feel that it tackles deeper, because more embracing, problems. However, the point was to illustrate the fact that metaphysical presuppositions may have a definite influence on the way in which scientific research proceeds. And that point I have, I think, established, even if you feel that my metaphysics has led me up the garden path. And, after all, I am a biologist; it is plants and animals that I'm interested in, not clever exercises in algebra or even chemistry. The garden path has its attractions for the likes of us, and all of us who want to understand living systems in their more complex and richer forms are fated to look like suckers to our colleagues who are content to make a quick (scientific) buck wherever they can build up a dead-sure pay-off.

Since I am an unaggressive character, and was living in an aggressively anti-metaphysical period, I chose not to expound publicly these philosophical views. An essay I wrote around 1928 on 'The

Vitalist-Mechanism Controversy and the Process of Abstraction' was never published. Instead I tried to put the Whiteheadian outlook to actual use in particular experimental situations. So biologists uninterested in metaphysics do not notice what lies behind—though they usually react as though they feel obscurely uneasy—and philosophers like Marjorie Grene may get so far as to conclude that I am not a wholly orthodox mechanical materialist.   [1969]

2

THE EVOLUTION OF DEVELOPMENTAL SYSTEMS

The theory of evolution is still the essential thread which connects together the numerous and diverse branches of biology. It has been clear at least since von Baer's day that a theory of evolution requires, as a fundamental part of it, some theory of development. Evolution is concerned with changes in animals, and it is impossible profitably to discuss changes in a system unless one has some picture of what the system is like. Since every aspect of an animal is a product of development, or rather is a temporary phase of a continuous process of development, a model of the nature of animal organization can only be given in developmental terms.

In recent times Goldschmidt has been the most prominent biologist who has attempted to describe biological organization in terms which are at once developmental and not too far removed from the genetical concepts employed by students of evolution. His great contribution to the topic was made by the publication in 1927 of his *Physiologische Theorie der Vererbung* [1]. He pointed out that the development of an animal consists of a large number of correlated reactions proceeding at definite relative velocities; and he suggested that genes act by altering the rates of one or more of the reactions. This fruitful idea was successful in directing the attention of many geneticists to development problems, and has been the stimulus to much valuable work. It in no way diminishes the historical importance of Goldschmidt's theory to point out, after this lapse of time, that it is actually no more than the statement of the general notion of materialism in a four-dimensional world. So long as one considers development in material terms, there are only two things a gene could do: alter the velocity of a reaction (which includes arresting it completely), or initiate a new reaction; and the second of these can always be looked on as a secondary consequence of an alteration in the rate of some earlier reaction.

Before the theory can give us any specific picture of the kind of thing a developing animal is, it must be elaborated and restricted in such a way as to give it some concrete consequences. Goldschmidt attempted to do this by specifying a particular mechanism for the alterations in rate, which he suggested were caused by alterations in the quantity of gene material, which was thought of as acting in an enzyme-like way. Interesting as this suggestion is, it is not quite to the point so far as evolution is concerned. What evolutionary theory requires is not so much a hypothesis of the ultimate physico-chemical mechanisms of development, but rather a picture of the possible kinds of interactions between developmental processes; it needs some framework of concepts in which it can discuss such questions as whether there is any difference in kind between specific and varietal differences.

I have recently [2] made some suggestions for an elaboration of Goldschmidt's hypothesis, which, it is hoped, will leave it general enough to apply to most, or all, developing organisms, but not so general that it applies equally well to the whole material universe. The most important of these suggestions can be summarized as follows:

1. The course of a developmental reaction is the resultant of a large number of mutually interacting influences. If one plots a developmental process against time, one must regard the line as occupying a position of equilibrium, determined by a number of processes which tend to push it upwards and a number which tend to push it downwards. This notion might seem to be merely our old friend dialectical materialism rearing its head out of the shifting seas of metaphysical controversy. But it is more than that; it follows directly from the experimental data concerning the effects of modifying genes, and of the general genetic background. It also issues in the following generalization from experience.

2. In a normal animal, there are only a certain finite number of possible resultants of the interacting developmental processes. The final products of development, the adult tissues, do not, it seems, vary continuously from one type to another, but fall into a comparatively few tolerably sharply defined kinds. This implies that during development there is a succession of 'branching points', at each of which the course of development can move into one or other of a few alternative paths.

3. One of the difficulties of an embryological theory has been to find a characteristic of a tissue which is causally connected with its future development. The classical concept of a potency is unsatisfactory because it is non-causal; it merely allows one to describe whether or

not a tissue does sometimes develop in a certain way. One can discover a characteristic of the required kind during a 'branch point'. At that time a tissue is in a state of indeterminacy between alternatives. This state, which I have called competence, is open to experimental investigation. It has the character of a readiness to react to certain stimuli (evocators), which may be applied either externally or, probably, from inside the cell. The future character of development depends directly on whether or not such a reaction occurs.

4. The definiteness of the alternative modes of development is a product of natural selection, and characterizes animals in the state of Nature. It is lost, or partially obscured, in most mutant forms.

5. Other genes may cause particular regions of the body to enter upon a type of development which should characterize some other part. A clear example is the gene Aristopedia in *Drosophila*, which causes the arista (terminal section of the antenna) to develop in a leg-like manner.

Some, at any rate, of these points are adumbrated in Goldschmidt's writings; but they have not been clearly realized by him. In his recent Silliman lectures [3] he is still concerned to show that time is the one and only essence of the matter. For example, he discusses the case of Aristopedia, and quotes his student Braun that the imaginal bud of the Aristopedia antenna becomes determined to develop into a leg at the same time as the leg buds are determined; but he draws the completely unjustified conclusion that it is determined as a leg simply because of this temporal simultaneity. One might as well argue that because the neural plate and the epidermis of an amphibian become determined at the same time they must be determined to develop in the same way.

Goldschmidt's book is one of the most important of recent contributions to the theory of evolution. A very valuable feature of it is his masterly analysis of geographical variation, a subject in which his own researches have been of supreme importance. He presents considerable evidence that animals in Nature fall into groups between which there are 'unbridgeable gaps'. The groups may be small relict species or relatively enormous assemblages of intergrading forms. Their taxonomic treatment will be a matter of convenience; but whether they are lumped as one species, or split up into a series of types each dignified with a specific name, is of minor importance in comparison with the essential fact of the existence of discontinuities between one set of intergrading types and the next. Evolutionary change within the group Goldschmidt names micro-evolution, while the arising of a new group separated from the old by a discontinuity, is macro-evolution.

A somewhat similar distinction has recently been made in the botanical field by Willis [4]. He suggests that the large groups arise by the working of 'some definite law which we do not yet comprehend', and that then 'evolution goes on in what one may call the downward direction from family to variety'. Goldschmidt's view is somewhat similar in that he supposes that the evolution within the group merely brings about an increase in variability without producing any progressive evolutionary changes; but, unlike Willis, he allows that this micro-evolution is controlled by natural selection. His main criticism is directed, not at the theory of selection, but at the idea that macro-evolutionary change is dependent on genes. He is 'firmly convinced that, except in micro-evolution, the facts already available to-day force us to drop completely from evolutionary thought the idea of so-called gene mutation, whatever it turns out to be physically or chemically'.

It is remarkable to find that this very radical notion is advanced on almost exactly the same grounds as were used by the Mendelians of the beginning of this century to show that evolution *does* depend on genes. Goldschmidt is unduly preoccupied by the fact that the experimental and mathematical development of the theory of natural selection have been mainly concerned, for technical reasons, with the small-scale evolutionary differences which separate one variety from another of the same species. But it was exactly the existence of large discontinuities of variation which originally gave plausibility to the view that gene differences are important in Nature, and Goldschmidt brings forward no compelling new evidence for the rejection of this view.

One reason for Goldschmidt's distrust of genes as adequate building materials for species may be that his 'rate concept' of gene action allows no place for differences in kind in the organisms produced; alterations in rate are essentially continuously variable. On the basis of the somewhat fuller theory given above, qualitative differences can easily be envisaged. According to that theory, the causal structure of an animal can be represented as a set of branching developmental paths, along each of which a certain part of the egg moves during its development. Changes in such a system would be primarily of two kinds: firstly, changes in the pattern or topological relations of the paths, which would imply an alteration in the distribution of the different portions of the egg between the various tissues, and, secondly, alterations in the actual course of the paths, which would imply that the final adult tissues were changed; and there might be combinations of these two kinds of change. Alterations of the first kind, which are typified by such a mutant as Aristopedia, would, I think, be reckoned as producing differences in kind, although it is always a matter of definition whether a change is taken as quantitative or qualitative. Differences of this sort have certainly played a large part in evolution on the grand scale, which gives rise to the major phyla.

The 'systemic' variations of Goldschmidt, that is to say the differences which separate the groups which are isolated by 'unbridgeable gaps', may also be of the second kind mentioned above. An alteration in the course of a developmental path will, if it occurs early in development, shift the whole set of paths which afterwards branch from it; that is to say, there will be a change in the character of a large number of tissues. This is the kind of difference which one most frequently finds as an 'unbridgeable gap'; it is a difference not in a few crude respects, but a permeating change in general facies. One finds genes which have similar effects within a species; for example, dachsous and rotund in *Drosophila*.

The 'unbridgeable' nature of the gap produced by such a gene can be understood in the following way. A developmental path has to be equilibrated; natural selection must build up a genetic background which stabilizes each path at the optimum. If, by an early-acting gene, a whole set of paths are thrown out of their old equilibria, a very considerable modification of the genetic background will be called for. Once accomplished, this will not be easily reversed or copied; and the new form will be effectively isolated from the old.

Goldschmidt's theory as to the causation of systemic variations is very unorthodox. He suggests that changes in the general pattern of animal are caused by complex rearrangements of the chromatin. This may possibly be so, but it is a long shot in the dark. There is some evidence which seems to tell against both the essential features of such a hypothesis. In the first place some specific differences, such as that between *Drosophila melanogaster* and *D. simulans*, involve less chromosome rearrangement than may be found within a species. In the second place, there is no reliable evidence that the position effects caused by rearrangements produce more radical alterations of development than do normal gene mutations; in fact, most position effects are slight quantitative differences.   [1941]

3

CANALIZATION OF DEVELOPMENT
AND THE INHERITANCE OF ACQUIRED CHARACTERS

The battle, which raged for so long between the theories of evolution supported by geneticists on one hand and by naturalists on the other, has in recent years gone strongly in favour of the former. Few biologists now doubt that genetical investigation has revealed at any rate the most important categories of hereditary variation; and the classical 'naturalist' theory—the inheritance of acquired characters —has been very generally relegated to the background because, in the forms in which it has been put forward, it has required a type of hereditary variation for the existence of which there was no adequate evidence. The long popularity of the theory was based, not on any positive evidence for it, but on its usefulness in accounting for some of the most striking of the results of evolution. Naturalists cannot fail to be continually and deeply impressed by the adaptation of an organism to its surroundings and of the parts of the organism to each other. These adaptive characters are inherited and some explanation of this must be provided. If we are deprived of the hypothesis of the inheritance of the effects of use and disuse, we seem thrown back on an exclusive reliance on the natural selection of merely chance mutations. It is doubtful, however, whether even the most statistically minded geneticists are entirely satisfied that nothing more is involved than the sorting out of random mutations by the natural selective filter. It is the purpose of this short communication to suggest that recent views on the nature of the developmental process make it easier to understand how the genotypes of evolving organisms can respond to the environment in a more co-ordinated fashion.

It will be convenient to have in mind an actual example of the kind of difficulties in evolutionary theory with which we wish to deal. We may quote from Robson and Richards [1]: 'A single case will make the difficulty clear. Duerden [2] has shown that the sternal, alar, etc., callosities of the ostrich, which are undoubtedly related to the crouching position of the bird, appear in the embryo. The case is analogous to the thickening of the soles of the feet of the human embryo attributed by Darwin [3] "to the inherited effects of pressure". As Detlefsen [4] points out, this would have to be explained on selectionist grounds by the assumption that it was of advantage to have the callosities, as it were, preformed at the place at which they are required in the adult. But it is a large assumption that variations would arise at this place and nowhere else.'

In this case we have an adaptive character (the callosities) of a kind which it is known can be provoked by an environmental stimulus during a single lifetime (since skin very generally becomes calloused by continued friction) but which is in this case certainly inherited. The standard hypotheses which come in question are the two considered by Robson and Richards: the Lamarckian explanation in terms of the inheritance of the effects of use, which they cannot bring themselves to support at all strongly, and the 'selectionist' explanation, which, in the form in which they understand it, leaves entirely out of account the fact that callosities may be produced by an environmental stimulus and postulates the occurrence of a gene with the required developmental effect. A third possible type of explanation is to suppose that in earlier members of the evolutionary chain, the callosities were formed as responses to external friction, but that during the course of evolution the environmental stimulus has been superseded by an internal genetical factor. It is an explanation of this kind which will be advanced here.

The first step in the argument is one which will scarcely be denied but is perhaps often overlooked. The capacity to respond to an external stimulus by some developmental reaction, such as the formation of a callosity, must itself be under genetic control. There is little doubt, though no positive evidence in this particular case so far as I know, that individual ostriches differ genetically in the responsiveness of their skin to friction and pressure. If we suppose, then, that in the early ostrich ancestors callosities were formed by direct response to external pressure, there would be a natural selection among the birds for a genotype which gave an optimum response.

The next point to be put forward is the one which is, perhaps, new in such discussions, and which therefore requires the most careful scrutiny. It is best considered as one general thesis and one particular application of it.

The main thesis is that developmental reactions, *as they occur in organisms submitted to natural selection,* are in general canalized. That is to say, they are adjusted so as to bring about one definite end-result regardless of minor variations in conditions during the course of the reaction.

The evidence for this comes from two sides, the embryological and the genetical. In embryology we have abundant evidence of canalization on two scales. On the small scale of single tissues, one may direct attention to the obvious but not unimportant fact that animals are built up of sharply defined different tissues and not of masses of material which shade off gradually into one another. Similarly, from the experimental point of view, it is usual to find that, while it may be possible to steer a mass of developing tissue into one of a number of possible paths, it is difficult to persuade it to differentiate into something intermediate between two of the normal possibilities. Passing from the scale of tissues to that of organs, it is not too much to claim it as a general rule that there is some stage in every life-history (though it may be an extremely early and short stage) when minor variations in morphology become 'regulated' or regenerated; and that is, again, a tendency to produce the standard end-product. Of course neither of these types of canalization is absolute. Morpho-logical regulation may fail if the abnormalities are too great or occur too late in development; and intermediate types of tissue can occasionally be found, particularly in pathological conditions.

The limitations on canalization which are important for our present purposes can better be seen when the problem is viewed from the other, genetical, side. The canalization, or perhaps it would be better to call it the buffering, of the genotype is evidenced most clearly by constancy of the wild type. It is a very general observation to which little attention has been directed (but see [5,6,7]) that the wild type of an organism, that is to say, the form which occurs in Nature under the influence of natural selection, is much less variable in appearance than the majority of the mutant races. In *Drosophila* the phenomenon is extremely obvious; there is scarcely a mutant which is comparable in constancy with the wild type, and there are very large numbers whose variability, either in the frequency with which the gene becomes expressed at all or in the grade of expression, is so great that it presents a considerable technical difficulty. Yet the wild type is equally amazingly constant. If wild animals of almost any species are collected, they will usually be found 'as like as peas in a pod'. Variation there is, of course, but of an altogether lesser order than that between the different in-dividuals of a mutant type.

The constancy of the wild type must be taken as evidence of the buffering of the genotype against minor variations not only in the environment in which the animals developed but also in its genetic make-up. That is to say, the genotype can, as it were, absorb a certain amount of its own variation without exhibiting any alteration in development. Considerable stress has been laid in recent years on certain aspects of this buffering. Fisher[8] and many authors following him have discussed 'the evolution of dominance', by which the genotype comes to be able to produce the standard developmental effects even when certain genes have been replaced by others of less efficiency. Again, Stern[9] and Muller[10] directed attention to the phenomenon of 'dosage compensation', by which it comes about that a single dose of a sex-linked gene in the heterogametic sex has the same developmental effect as a double dose in the homogametic. These two processes are part of the larger phenomenon which we have called the canalization of development. This also includes other, at first sight unrelated, features of the genotypic control of development. For example, attention has been directed [11] to genes which cause certain regions of developing tissue to take an abnormal choice out of a range of alternative possible paths; Mather and de Winton[12] have recently spoken of such genes as 'switch genes'. Finally, Goldschmidt has shown that environmental stimuli may, by switching development into a path which is usually only followed under the influence of some particular gene, produce what he has called a 'phenocopy' of a previously known mutant type.

There seems, then, to be a considerable amount of evidence from a number of sides that development is canalized in the naturally selected animal. At the same time, it is clear that this canalization is not a necessary characteristic of all organic development, since it breaks down in mutants, which may be extremely variable, and in pathological conditions, when abnormal types of tissue may be produced. It seems, then, that the canalization is a feature of the system which is built up by natural selection; and it is not difficult to see its advantages, since it ensures the production of the normal, that is, optimal, type in the face of the unavoidable hazards of existence.

The particular application of this general thesis which we require in connexion with 'the inheritance of acquired characters' is that a similar canalization will occur when natural selection favours some characteristic in the development of which the environment plays an important part. It is first necessary to point out the ways in which the environment can influence the developmental system. If we conceptually rigidify such a system into a definite formal scheme, we can think of it as a set of alternative canalized paths; and the environment can act either as a switch, or as a factor involved in the system of mutually interacting processes to which the buffering of the paths is due. This is, of course, too dead and formal a scheme to be a true picture of development as it actually occurs. In so far as it is always to some extent, but not entirely, a matter of convenience what we decide to call a complete organ, so far will it be a matter of convenience what we consider to be different alternative paths; and the question of whether a given influence is thought of as a switch mechanism or a modification of a path will depend on how we choose our alternatives. There are some cases, however, in which the alternatives are very clearly defined. Thus it is commonly assumed that the evolution of sexuality passed through a stage in which, as in *Bonellia*, the environment acted as a switch between two well-defined alternatives; later, genetic factors arose which superseded the environmental determination by an internal one.

More commonly, however, the original environmental effect will be to produce a modification of an already existent developmental path. Thus in the case of the ostrich ancestors, the formation of callosities following environmental stimulation is a response by a developmental system which is normally present in vertebrates. This system must, in all species, be subject to natural selection; outside certain limits, too great or too low a reactivity of the skin would be manifestly disadvantageous. If we suppose that the callosities, when they were first evolved, were dependent upon the environmental stimulus, then the evolution appears as a readjustment of the reactivity of the skin to such a degree that a just sufficient thickening is produced with the normally occurring stimulus.

There would appear to be two possible ways in which such a development might be organized. It might on one hand remain uncanalized, the formation of the thickening in each individual depending on the reception of the adequate stimulus, to which the response remained strictly proportional. If this possibility was realized, the well-known difficulty of accounting for the hereditary fixation of the character remains unimpaired. The alternative is that the development does become canalized, to a greater or lesser extent. In that case, the magnitude of the response would not be proportional to that of the stimulus; there would be a threshold of stimulus, above which the optimum (that is, naturally selected) response would be formed. In so far as the response became canalized, the environment would be acting as a switch.

Systems of either type can be built up by natural selection, and one can point to examples of them in animals at the present day. The reaction of the patterns on Lepidopteran wings (for example, in *Ephestia*[13]) to temperature during the sensitive period scarcely seems to involve thresholds, while the metamorphosis of the axolotl, for example, clearly does. In general, it seems likely that the optimum response to the environment will involve both some degree of proportionality and some restriction of this by canalization. The most favourable mixture of the two tendencies will presumably differ for different characters. It is easy to see why a much sharper distinction between alternatives is generally evolved in connexion with sex differences than with the degree of muscular development, for example; but even the former is to some extent modifiable by extreme and specialized environmental disturbances (heavy and early hormone treatment), and even the latter has some degree of genetic determination.

The canalization of an environmentally induced character is accounted for if it is an advantage for the adult animal to have some optimum degree of development of the character irrespective of the exact extent of stimulus which it has met in its early life; if, for example, it is an advantage to the young ostrich going out into the hard world to have adequate callosities even if it were reared in a particularly soft and cosy nest. Now in so far as the development of the character becomes canalized, the action of the external stimulus is reduced to that of a switch mechanism, simply in order that the optimum response shall be regularly produced. But switch mechanisms may notoriously be set off by any of a number of factors. The choice between the alternative developmental pathways open to gastrula ectoderm, for example, may be made by the normal evocator or by a number of other things (the mode of action of which may be through the release of the normal evocator (cf. [14]), but which remain different to the normal evocator nevertheless). Again, we know many instances in which several different genes, by switching development into the same path, produce similar effects; and attention has already been directed to the 'phenocopying' of a gene by a suitable environmental stimulus. Thus once a developmental response to an environmental stimulus has become canalized, it should not be too difficult to switch development into that track by mechanisms other than the original external stimulus, for example, by the internal mechanism of a genetic factor; and, as the canalization will only have been built up by natural selection if there is an advantage in the regular production of the optimum response, there will be a selective value in such a supersession of the environment by the even more regularly acting gene. Such a gene must always act before the normal time at which the environmental stimulus was applied, otherwise its work would already be done for it, and it could have no appreciable selective advantage.

Summarizing, then, we may say that the occurrence of an adaptive response to an environmental stimulus depends on the selection of a suitable genetically controlled reactivity in the organism. If it is an advantage, as it usually seems to be for developmental mechanisms, that the response should attain an optimum value more or less independently of the intensity of stimulus received by a particular animal, then the reactivity will become canalized, again under the influence of natural selection. Once the developmental path has been canalized, it is to be expected that many different agents, including a number of mutations available in the germplasm of the species, will be able to switch development into it; and the same considerations which render the canalization advantageous will favour the supersession of the environmental stimulus by a genetic one. By such a series of steps, then, it is possible that an adaptive response can be fixed without waiting for the occurrence of a mutation which, in the original genetic background, mimics the response well enough to enjoy a selective advantage.   [1941]

4

SELECTION OF THE GENETIC BASIS
FOR AN ACQUIRED CHARACTER

There are so many examples of the adaptation of an animal to its environment which at first sight would appear to find their simplest explanation in the supposition that the effects of the environment have become inherited, that theories of this kind have continued to retain a following in spite of the lack of clear experimental evidence in their support. This following has been composed mainly of naturalists; experimentalists and geneticists have recently tended to adopt an attitude similar to that expressed by Dobzhansky[1], who writes: 'This question has been discussed almost *ad nauseam* in the old biological literature . . . so that we may refrain from the discussion of it altogether'. In dismissing the matter so cavalierly, Dobzhansky was explicitly referring to 'direct adaptation', that is, the hypothesis that when the environment produces an alteration in the development of an animal, it simultaneously causes a change in its hereditary qualities such that the development alteration tends to be inherited. It has been usual, indeed, to consider this suggestion as the only possible alternative to the opposed view that environmental effects have no hereditary consequences, the phenomena of adaptation being solely due to the natural selection of chance variants.

Recent work, however, suggests another alternative. We know that environmental stimuli may produce developmental abnormalities (phenocopies) which simulate the effects of mutant genes; and a perusal of the scattered literature on the subject suggests that there is a good deal of variation among normal stocks in their sensitivity to the external stimuli. I therefore suggested some years ago[2] that all the necessary machinery is available by which a genetic basis could be, and in the course of natural selection would be, set up to reproduce any given environmental effect which was of value to the animal concerned. Natural selection will act, not solely on fortuitous variants resembling the form produced by the environment, but on the sensitivity of normal individuals to the environmental stimulus; and the genotypes sensitive to the external influence will also reinforce the action of any genes which tend to produce similar phenotypes and will canalize their activity towards the exact effect which is being selected for.

In a recent experiment, this possibility has been actually realized. Individuals of a wild-type strain of *Drosophila melanogaster* were given a strong environmental stimulus by submission to a temperature of 40°C for four hours at an age of 17–23 hours (in later generations 21–23 hours) after puparium formation. A crossveinless phenocopy was produced with a frequency of about 40 per cent. Two selection lines were set up, in one of which the flies which showed the phenocopy were bred from in each generation, while, in the other, selection was against phenocopy formation. Fairly rapid changes were

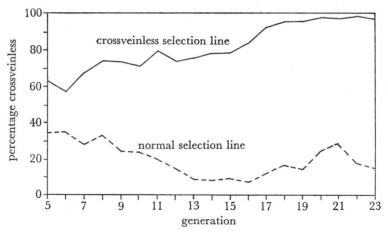

Figure 1. Progress of selection for and against the formation of the crossveinless phenocopy, from the fifth generation onwards, the temperature shock being applied to pupae aged 21–23 hr.

produced in both directions, particularly from the fifth generation onwards, at which time it was realized that the critical period for the effect was at the 21–23 hour stage, to which the treatment was thereafter confined (see figure 1).

The important point is that from the twelfth generation onwards of the stock selected for sensitivity, flies showing the crossveinless phenotype began to appear even among those individuals which had not been given the temperature treatment. Matings between these have given rise to strains which regularly produce cross-veinless flies when cultured at normal temperatures. During the course of selection, a genetic constitution has therefore been synthesized which under normal conditions produces the same effect as was originally found only as a response to the stimulus of an abnormal environment. This genetic constitution is not, in the present stocks, fully penetrant, the frequency of crossveinless in the various strains reared at normal temperature never surpassing about 80 per cent; but there is no reason to suppose that further selection will not elicit a fully penetrant strain.   [1952]

5

THE EVOLUTION OF ADAPTATIONS

It is abundantly found in the living world that the structure of an animal or plant is very precisely adapted to the functions by which it has to perform. The nature of the processes by which this situation has been brought about during evolution provides one of the major problems for biological theory. The hypothesis of the inheritance of acquired characters suggested that in some way or other the effects of functioning become themselves inherited. It has usually been interpreted to mean that the reaction between the organism and its surroundings has, as one of its results, an effect on the germ-plasm such that new hereditary changes occur, of a kind which determines the development in later generations of individuals suited to these particular conditions of life. Although this idea has recently been revived in a rather nebulous form in the Soviet Union, it has been so completely rejected by the rest of the scientific world that it is hardly considered to be worthy of discussion in most of the important recent works on evolution. The reigning modern view is that, in nature, the direction of mutational change is entirely at random, and that adaptation results solely from the natural selection of mutations which happen to give rise to individuals with suitable characteristics. I want to argue that this theory is an extremist one, and that, in essaying to account for adaptation, it neglects to call to its aid the doctrines emerging in other fields of modern biology which can quite properly be combined with the conclusions of genetics in the strict sense. In the discussion which follows, attention will be confined to animals, but there is no reason to doubt that similar arguments could be advanced in the botanical field.

It will be advisable first to glance briefly at the phenomena which are usually referred to under the heading of adaptation, since they are of several different kinds which must be distinguished from one another.

There is, first, a category in which an animal living under particular circumstances, or behaving in a particular way, itself becomes modified so as to be better fitted for its special circumstances. Examples of such 'exogenous' adaptations are legion. If muscles are continually and intensely used, they become thicker and stronger; if one kidney is removed from a mammal, the other hypertrophies; if the forelegs are absent at birth, or removed shortly afterwards, from rats or dogs, the hind-limbs become modified to suit the bipedal gait which the animals are forced to adopt; if skin is subjected to frequent rubbing and pressing, it thickens and becomes more horny; and one could multiply such instances almost indefinitely.

Secondly, there is a category of what may be called pseudo-exogenous adaptations, in which the animal exhibits characteristics similar to effects which can be called forth as direct exogenous adaptations, but which on investigation are shown to be hereditary, and independent of any particular environmental influence. We shall consider some examples of such adaptation in more detail later, since they pose one of the most striking problems to be solved.

Finally, there is a very large class of adaptations, which, as Medawar [1] has recently emphasized, need to be distinguished from the previous category, and which are characterized by the fact that the adaptive feature is of a kind which one cannot imagine as having ever been produced in direct response to the environmental conditions or mode of life of the animal. To give two examples only, Medawar mentions the modifications of certain epidermal cells to secrete sweat, and the development, from another part of the skin, of a transparent area which forms the cornea of the eye. It is, as he says, impossible to see how any attempt to peer through an area of opaque skin could tend to cause it to become transparent. The adaptation of the cornea to vision can hardly have arisen in any causal dependence on external factors, and we might therefore give the name of endogenous adaptations to this category.

It is in connection with this third type of adaptation that we can as yet make the least progress beyond the current hypothesis, which is content to rely on the chance occurrence of suitable mutations. But even here there is a little more to be said. In some endogenous adaptations, the usefulness of the character is in connection with factors in the outside world. Another of Medawar's examples, the possession of horns which serve the purposes of aggression or defence, will suffice as an instance. But the transparency of the cornea is adaptive because it is suited to the functioning of another internal part of the organism, namely the retina, sensitive to the light which the cornea allows to enter. Now, I think that we shall often find that the various parts concerned in such internal endogenous adaptations are involved with one another not only during their functioning in the adult animal, but during their development in the embryo. This is certainly true of the cornea, which can be induced from normal epidermis if an eye-cup is transplanted under it at an early enough stage. All the various parts of the eye, which can function efficiently only if they have the correct relations with one another, are interdependent during their development. It was shown many years ago by Ross Harrison[2] that if the large lens of the axolotl *Amblystoma tigrinum* is grafted over the eye-cup of the smaller *A. punctatum*, the lens does not grow to its full size, while the eye-cup provided with the larger lens attains a greater size than usual (figure 1). Similarly, if the eye-cup of *A. tigrinum* is transplanted under the early embryonic skin of *A. punctatum*, it induces a lens which is originally of *A. punctatum* size and therefore relatively too small, but, as growth proceeds, the lens grows faster and the eye-cup more slowly than usual, so that they gradually achieve the normal relative proportions. We might say that the internal adaptation of the lens to its retina, although endogenous in the sense that it arises within the animal, is nevertheless affected by factors from outside the lens itself, namely by the retina. The problem it presents is therefore not wholly different from that offered by adaptations to external factors. Moreover, the adaptation is in part a direct response to the influence of the retina, and is thus similar to the first category distinguished above; but there is clearly also some inherent tendency for the *A. tigrinum* material to grow faster than the *A. punctatum*, and in this respect we are reminded of the second category, of pseudo-exogenous adaptations.

We may next consider the true exogenous adaptations, in which an animal becomes modified by external factors in such a way as to increase its efficiency in dealing with them. Such adaptation does

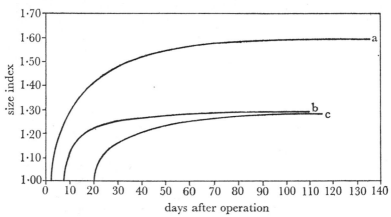

Figure 1. Transplantation of lens ectoderm from *Amblystoma tigrinum* to *A. punctatum*. Curve (a): growth of the *tigrinum* lens when associated with the eye-cup of the same species, expressed as the ratio of the size of *tigrinum* lens to the size of *punctatum* lens. Curve (b): similar curve for *tigrinum* lens associated with *punctatum* eye-cup. Curve (c): growth of *punctatum* eye-cup associated with *tigrinum* lens, expressed in terms of its size when associated with *punctatum* lens. Note how in the interspecific combination the lens grows more slowly and the eye-cup faster than they would normally do (after Harrison).

not, of course, always occur: the environment may overcome the animal and eventually kill it. For instance, specialized breeds of cattle from a temperate zone growing up in the tropics may fail to become acclimatized, and, instead, develop into degenerate forms which have great difficulty in surviving. Even in such cases, however, there are probably always some tendencies towards the development of useful adaptations, though they are not strong enough to be effective. Adaptation to relatively slight changes in environment is, however, usually successful. We still know comparatively little as to how this is brought about, although we can speculate more or less plausibly about the mechanisms involved in particular cases. The hypertrophy of one organ of a pair when the other is removed (as in the case of kidneys) may, for instance, find its explanation in Haddow's suggestion [3] that each organ requires some special specific substance for its growth, or in Weiss's interesting work [4] on the effects on the growth of an organ of antisera specific against it (see also [5]).

It may be that, in many cases, the chemical processes involved in the functioning of an enzyme or similar substance are related to those by which it is synthesized, so that increased function automatically means increased synthesis. Other examples are certainly more complex. For instance, if newt larvae are kept in water poor in oxygen, not only do the gills grow larger, but the tissue in their walls becomes thinner, allowing a more rapid diffusion of gas[6]. Possibly here we might be able to find some explanation in terms of the effect of oxygen-lack on the rate of blood-flow, and the influence of that in its turn on the vessels through which it is moving. But the phenomenon of adaptive response is so widespread and so general that it seems hardly convincing to explain it only by a series of *ad hoc* hypotheses, invoking a new one for each different case. At present, we hardly seem able to do better than go to the other extreme, and produce the general argument—too general to be very satisfying— that it is an advantage to animals to be able to adapt successfully to new circumstances, and therefore natural selection will have favoured those animals which by chance had a hereditary endowment which enabled them to do so. One feels that, somewhere between these extremes of particularity and inclusiveness, there should be principles of fairly general application to be discovered—but they still await their Darwin.

It is the remaining category of adaptations, the pseudo-exogenous, which has provoked the most discussion. We are confronted here by phenomena for which an explanation could so easily be found in a direct effect of some environmental factor, were it not that further study demonstrates unequivocally that the structure concerned is determined by the heredity of the organism, and is relatively independent of the environment. The question arises whether we can bring ourselves to believe that the part which the environment can play in mimicking the condition is really irrelevant, and that the evolution of this particular adaptation has resulted from the selection of chance mutations which might have appeared and produced the phenotypes even if the environmental effects had never existed.

Some concrete examples will make the problem clearer. One of the most familiar is that of the thickened skin on the soles of our feet. This thickening is obviously an adaptation to the stresses which this region of the body has to bear; but, as Darwin pointed out, and as Semon[7] discussed in a full-length paper, the thickening already appears in the embryo, before the foot has ever borne any weight. The structure therefore cannot be a direct response to external pressure, but must be produced by the hereditary constitution independently of the specific external influence to which it is an adaptation. The situation is even more striking when similar thickenings are found on less conventional parts of the body. For instance, the ostrich squats down in such a way that the under surface of the body comes into contact with the ground at its two ends, fore and aft. In just these places a considerable callosity develops in the skin (figure 2), and Duerden[8] showed that these thickenings make their appearance in the embryo before hatching. The same thing is true of callosities which appear on the wrists of the forelegs of the African wart-hog, which while feeding has a peculiar stance which involves resting on these points[9]. Another remarkably clear example, affecting a different organ, is that of the second molar tooth in the dugong. In the adult, the crown of this tooth is more or less flat, with slight transverse ridges. This shape

Figure 2. The underside of an ostrich's body, showing the two callosities (drawn from a photograph published by Duerden).

must be regarded as a modification of an originally more conical tooth, and would at first sight appear to be directly related to the use of the tooth for grinding the food. However, Kükenthal[10] showed that, although the tooth first appears in a conical form in the early embryo, processes of resorption begin to change this into the final flattened form before the tooth is used (figure 3).

 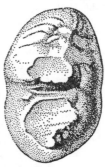

Figure 3. Two stages in the development of the second upper molar of the dugong. In the young embryonic stage on the left, the tooth has three conical protuberances, in which resorption is just beginning. In the older but still embryonic stage on the right, more or less flat faces have appeared (after Kukenthal).

It certainly seems very far-fetched to attempt to explain such phenomena without bringing in the fact that the environment might be expected to produce similar effects. Let us consider, therefore, what might happen to an ostrich in which the appropriate callosities were not hereditarily determined. Presumably its skin, like that of most other animals, would react directly to external pressure and rubbing by becoming thicker. Now the point which seems to have been overlooked in previous discussions of the matter is that this capacity to react must itself be dependent on genes. Since populations of animals are never quite uniform in any character, we must expect that the ostrich ancestors varied in their capacity to produce the most suitable callosities; and there could be effective natural selection for those which performed the most satisfactory exogenous adaptation. A race would evolve in which the stresses set up by squatting in a particular way would call forth the development of appropriate adaptive thickenings of the skin.

At this stage, the thickenings would still not be hereditary and independent of the pressure and rubbing; they would still be acquired characters in the conventional sense. We can find a hypothesis of how they might come to be hereditarily fixed if we turn to consider another aspect of the matter. The callosities are the results of developmental processes. Now, one of the main characteristics of animal development is that it tends to be canalized or buffered, so that the optimum end-result is produced even if there are minor variations from the normal conditions while the process is going on [11]. Natural selection, in fact, does not merely ensure that only those animals survive which have something near the optimum characteristics, but favours those genotypes which tend to produce such animals under any conditions. It gradually builds up efficient cybernetic mechanisms, to use a fashionable phrase. Thus we may expect to reach a stage in which our ostriches nearly always develop callosities of just the right size and position, even in those individuals which, to put it crudely, sit down very seldom or those which loll about the whole time.

Once such a cybernetic developmental mechanism has been built up, it will be rather like a gun set to go off when the trigger is pulled. The development of the callosities will proceed quite autonomously, once the process can be started. The initial stimulus, which may be greater or lesser amount of external pressure, has become a relatively minor factor in the whole situation. It may then not be too difficult for a gene mutation to occur which will modify some other nearby region of the embryo in such a way that it takes over the function of the external pressure, interacting with the skin so as to 'pull the trigger' and set off the development of the callosities [12].

This may have seemed too long a train of argument to be very convincing, but a good deal of experimental evidence can now be produced to support it. In the first place, there seems to be no doubt that the formation of callosities on the sole of the foot is actually brought about by developmental interactions which go on in the embryonic limb-bud, as was suggested in the last paragraph. In animals which are hereditarily polydactylous, the genetic constitution causes an increase in the number not only of skeletal elements, but of the associated structures such as muscles, and, in particular, callosities (figure 4). These thickenings are therefore an integral part of the presumably complex but as yet unanalysed actions and reactions which mould the early limb-bud into the adult foot.

A more comprehensive check on the whole train of thought may be obtained if we imitate the postulated action of natural selection by artificially selecting for the ability to respond in some definite way to an environmental stimulus. No such work has been done on

Figure 4. Right hind foot of a normal guinea-pig (above), and of a polydactylous guinea-pig (after S. Wright).

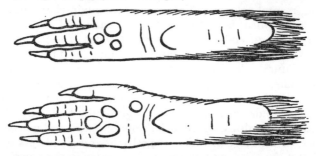

the formation of callosities, but an experiment has been made on the rapidly breeding fruit-fly *Drosophila melanogaster* [14, 15]. In order to make the situation as clear-cut as possible, an environmental stimulus was used which produced a phenotype which did not normally occur at all in the stock used. If pupae aged about 21–23 hours are subjected to a temperature of 40°C for four hours, a proportion of them develop wings in which the posterior cross-vein (and occasionally the anterior one) is broken or missing (figure 5).

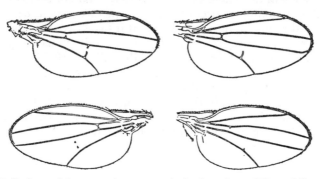

Figure 5. Defects of the posterior cross-vein in the wings of *Drosophila melanogaster* which had been given a temperature shock in early pupal life.

Two selection-lines were set up, in one of which only flies with broken cross-veins were used as parents; in the other, the flies which failed to react were selected. In both lines, the percentages of reactive individuals changed as time went on, increasing in the upward selected line and decreasing in the other. It is therefore actually possible to select for the capacity to respond to the environment (figure 6).

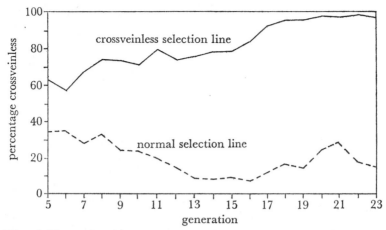

Figure 6. The results of selection for and against the ability to respond to a temperature shock by the formation of a cross-veinless phenotype.

As the experiment proceeded, it confirmed further points in the theory given above. Stabilization of the type with broken or missing cross-veins did in fact take place, and proceeded so far that, as was predicted, strains could be produced which lacked the cross-vein even when they had been reared for their whole life under the standard normal conditions, and had never been subjected to the high-temperature treatment of their pupae. The cross-vein effect, which in the original stock had appeared only as a response to an external stimulus, is in these selected races quite independent of any special feature of the environment. We may say that the acquired character has been 'assimilated' by the genotype. Presumably, the lack of the vein is due now to some modification of the reactions which go on autonomously within the developing wing, but we do not know enough about the physiology of wing development in *Drosophila* to guess exactly what the reaction might be.

The gradual increase in the frequency of cross-veinless flies shows quite definitely that selection has been at work. It may, however, still be asked whether all that happened was the concentration of genic variants which were present in the population to begin with, or whether new mutations tending to break the cross-vein have arisen during the course of the selection, and, if so, whether we can suppose that the treatment has itself caused them to appear. The experiments which have been made so far do not allow of a final answer. It may be pointed out that the selected cross-veinless lines differ from the foundation stock in quite a large number of genes; we are not dealing with only a single gene, and if what we have collected are new mutations, there must have been a large number of them. Moreover, a considerable number of flies, in the order of a thousand, were involved in each generation, and it seems unlikely that new mutations could have been sorted out from such numbers in the comparatively few generations of the experiment. One cannot, however, absolutely rule out the possibility that the treatment provoked appropriate mutations. The main point is that there is no need to make this hypothesis, which is contrary to everything that we know about the mutation process. Selection of already existing genes which affect responsiveness to the environment, and stabilization of their effect, together provide a plausible account of the result in terms of orthodox genetic and embryological mechanisms.

It is probable that this process of genetic assimilation of an environmental effect has played a very important part in evolution. Pseudo-exogenous adaptations are very common in the animal, and also in the plant, world. Instances such as the ostrich's callosities and the dugong's tooth, which were discussed above, were chosen as particularly striking and impressive examples, not because they are typical; usually such adaptations have a much more ordinary character. One such case, which will also serve to illustrate some of the problems which remain to be cleared up, is that of the mountain forms of the common water snail *Limnaea peregra*[16]. In certain tarns and loughs in Scotland and Ireland having very soft water, this snail occurs with an unusually thin shell, a form to which the name *L. praetenuis* has been applied. The thinness of the shell is clearly an effect of the lack of calcium in the water. It is found that when these forms are bred in water of normal hardness, they revert more or less completely to the normal; the effect in this case therefore is entirely, or at least mainly, an exogenous one which has not been genetically assimilated. But there is another method of economizing in shell-forming materials, which involves coiling the shell more nearly in one plane, so that the spire is reduced. Some lakes contain races which have adopted this shape, usually combined with a certain thinning of the shell; they have been given the varietal names *burnetti* and *involuta*. When they are bred in hard water, many of the races revert to normal, but in some of them the involute character is persistent, and must have been genetically assimilated. A whole series of questions suggest themselves. Why is the *praetenuis* form apparently never genetically fixed, while the *involuta* form sometimes is? Is it because the direct action of the environment is more effective in producing a thin shell of normal shape than a shell of altered configuration? Or is the thinness of the shell in *praetenuis* not truly an adaptation which is advantageous to the animal, but merely a necessary consequence of the lack of calcium, while the altered shell form of *involuta* has a selective advantage? And why has involution been genetically assimilated in some races but not in others? Possibly the length of time for which the lake has been colonized, or the degree of involution attained, may be important in this connection, but we still know too little to say.

It is clear that the theory of genetic assimilation may have wide applications, but before it can be used with confidence it requires much more experimental support than it yet has. The first thorough experiment made to test it gave the expected result, but that is hardly sufficient. Attempts are now being made with *Drosophila* to select for the capacity to give other responses to various environmental stimuli. In some cases, the experience with the cross-veinless character seems to be being repeated, but in others there has been little reponse to selection over the first few generations, and it may turn out that the original populations have little available variation in capacity to respond in the necessary way, in which case genetic assimilation would have no chance to operate. Until such limitations on the process have been more fully worked out, some caution is called for in applying the theory to all the phenomena for which it seems able to provide an explanation. However, it is at least not unimportant that we now have a hypothesis which gives a more or less plausible explanation of this large class of adaptations without invoking the discredited theory of the direct inheritance of acquired characters, and in a more convincing way than by a mere reliance on the occurrence of a suitable chance mutation.　　　　　[1953]

EVOLUTION AND EPISTEMOLOGY

Prof. E. Schroedinger's recent book[1] discussing the relevance of ancient Greek thought to present-day scientific philosophy touches on very many of the problems which have been debated thoughout the whole history of Western civilization. He closes it (p. 94) with the thought, adumbrated perhaps already by Democritus (see p. 30), that 'for the purpose of constructing the picture of the external world, we have used the greatly simplifying device of cutting our own personality out, removing it ... this is the reason why the scientific world-view contains of itself no ethical values, no aesthetic values, not a word about our own ultimate scope or destination, and no God, if you please'.

But has not recent biological thought added to the picture a factor which Schroedinger fails to take into account? We believe now that the intellectual-sensory apparatus by which we make contact with the external world arose through a process of evolution. According to classical neo-Darwinian views, the process by which it came into being, and acquired the characteristics which it has in man, depended on the occurrence of random changes which were selecting according to their effectiveness in furthering the reproductive ability of the individual animals concerned. If this is true, it is no longer adequate to frame a discussion in terms of a dichotomy between 'our personality' and an impersonal world which we arrive at by 'cutting our personality out'. The faculties by which we arrive at a world view have been selected so as to be, at least, efficient in dealing with other existents. They may, in Kantian terms, not give us direct contact with the thing-in-itself, but they have been moulded by things-in-themselves so as to be competent in coping with them.

I believe one can go further than this in asserting a congruity between our apparatus for acquiring knowledge and the nature of the things to be known. I have suggested[2] that evolution operates not so much by the differential reproduction of chance variations of phenotype, but by the selection of genotypes which endow their possessors with the capacity to react adaptively with their surroundings (adaptive being defined in terms of reproduction). From this point of view it is easier to envisage the possibility that our mental equipment is not merely one that happens to be reasonably efficient, but that it has been shaped precisely to fit the character of those things with which it has to make contact. If this were so, the fundamentals of 'our personality' must still be implicit in the scientific world-view, and attempts to find them there (such as a tentative one of my own[3]) cannot be dismissed as impossible of fulfilment.

[1954]

7

EVOLUTIONARY ADAPTATION

The subject of this paper is the origin of adaptation, still an issue one hundred years after Darwin, and recently characterized by George Gaylord Simpson[1] as the 'primary problem of evolutionary biology'.

The assemblage of life-sciences that are usually classed together as 'biology' form a group at least as complex and diversified as the whole group of the physical sciences. Within this enormous range one can discern three main foci toward which the individual sciences tend to be oriented. One of these is analytical biology—the attempt to determine the ultimate constituent units upon which the character of living things depend. Analytical biology investigates development and heredity through the analysis of genes and subgenic units and so to the macrochemical entities such as the DNA, RNA, and protein of the chromosomes. The whole group of such studies plays the same role in the biological field as does atomic physics in the physical sciences.

Another major focus of interest is what may be called 'physiological biology'—the study of the mechanisms by which organisms carry on their existence. This corresponds perhaps to chemistry and engineering in the physical role.

Finally, there is what may be called 'synthetic biology', which is concerned with providing an intellectually coherent picture of the whole realm of living matter. In the structure of biology this fulfils the same role as cosmology does in the physical sciences; and just as in the physical realm we find that cosmology and atomic physics have very close connections with one another, so in biology the analytical and synthetic approaches to the world of the living employ very similar concepts.

In the hundred years since Darwin wrote, it has become universally accepted that the only synthetic biological theory which needs serious consideration is that of evolution. In an appraisal of evolutionary biology as it stands now in this centenary year, it is perhaps well to begin by reminding ourselves of the fundamental reasons for mankind's interest in this subject. During the century of intensive work which has been devoted to its study, so many detailed problems have emerged which have a great fascination of their own that one is sometimes inclined to be carried away by enthusiasm for these puzzles; however, they are really attractive only to those who have already taken their first steps toward this direction of study. The enormous impact of Darwin's theories on the whole intellectual life of his own day—and, indeed, on that of all later generations—arose not from details but from the relevance of the broad outline of his thinking to one of the major problems with which mankind is faced.

That problem is presented by the appearance of design in the organic world. Animals and plants in their innumerable variety present, of course, many odd, striking, and even beautiful features, which can raise feelings of surprise and delight in the observer. But over and above this, a very large number of them give the appear-

ance of being astonishingly well tailored to fit precisely into the
requirements which will be made of them by their mode of existence.
Fish are admirably designed for swimming, birds for flying, horses
for running, snakes for creeping, and so on, and the correspondence
between what an organism will do and the way it is formed to carry
out such tasks often extends into extraordinary detail.

It is clear from the oldest literatures that man has always been impressed by this correspondence. The simplest explanation—and the
one almost universally accepted in prescientific times—is that this
appearance reflects the activities of an intelligent Being who has designed each type of animal and plant in a way suitable for carrying
out the functions assigned to it. It is the challenge presented to this
explanation that constitutes the major interest of the theory of
evolution. A really convincing alternative account of the origin of
biological adaptation is the major demand which must be made of it.

The essential feature of an evolutionary theory is the suggestion
that animals and plants, as we see them exhibiting an apparently
designed adaptedness at the present day, have been brought to their
present condition by a process extending through time and were not
designed in their modern form. This does not, as many of Darwin's
contemporaries thought it did, necessarily deny the existence of any
form of intelligent designer. It means only that any designing activity
there may be has operated through a process extending over long
periods of time and has not brought suddenly into being each of the
biological forms as we now see them. The question of theism or
atheism, which played such a large part in the public discussions of
Darwin's day, is, we now recognize, not critically answered by the
acceptance or rejection of an evolutionary hypothesis but must be
settled—if it ever can be—in some other way. We need not, therefore, be further concerned with it in this discussion.

Evolutionary theories had, of course, been put forward some time
before Darwin wrote *Origin of Species*. The most famous of these
earlier discussions is that associated with the name of Lamarck. It has
suffered a most surprising fate. Lamarck is the only major figure in
the history of biology whose name has become, to all intents and purposes, a term of abuse. Most scientists' contributions are fated to be
outgrown, but very few authors have written works which, two cen-
turies later, are still rejected with an indignation so intense that the
sceptic may suspect something akin to an uneasy conscience. In point
of fact, Lamarck has, I think, been somewhat unfairly judged.

Lamarck's theory involved two main parts, and each of these has
encountered some essentially spurious difficulties in gaining accept-

ance. The first part supposed that the initial step towards an evolutionary advance involves something which Lamarck characterized as an act of will. Clearly, in this form the postulate applies only to animals and not to plants. Lamarck was, I take it, suggesting that the organism's own behaviour is involved in determining the nature of the environmental situation in which it will develop and to which its offspring will become adapted. In this form his theory could perhaps be generalized to cover the plant kingdom also if one accepts a wide enough definition of the concept of behaviour. However, let us leave that on one side: Lamarck himself was concerned primarily with animal evolution.

Now a concept such as an act of will was for a long time very unfashionable in the scientific study of biology. It is only relatively recently that biologists have shown any confidence in tackling the problems presented by the study of animal behaviour. Most students of behaviour still avoid such terms as 'act of will', but the concept of a choice between alternative modes of behaviour or conditions of life is by now quite respectable, and one must make allowances for the terminology used by someone writing in the eighteenth century.

If a certain sympathy is shown in interpreting Lamarck's words, the second phase of his theory also appears less unacceptable than it is usually considered to be. This is the well-known hypothesis of the inheritance of acquired characters. Conventionally, at the present time this is interpreted as though Lamarck used the word 'inheritance' as we should now use it, that is to say, to mean transmission of a character from a pair of parents to their offspring in the next or immediately subsequent generation. But, at the time Lamarck wrote, no distinction had yet been made between heredity over one or two generations, as we study it in genetical experiments, and heredity over much longer periods of time, as we encounter it in evolution. Nor was there any discrimination between the genetics of individuals and what we now call 'population genetics'. Lamarck's theory could quite well be interpreted to mean not that an individual organism which acquires a character during its lifetime will tend to transmit this to its immediate offspring, but that, if members of a population of animals undergoing evolution in nature acquire a character during their lifetime, this character will tend to appear more frequently in members of a derived population many generations later. In this form it is not so easy to reject his view. In fact, in a later part of this lecture I shall produce some evidence in favour of it.

Lamarck's words were, however, not interpreted in the way that I have suggested. His postulated 'act of will' was rejected as something vitalistic and non-scientific. His doctrine of the inheritance of acquired characters was interpreted in terms of individual genetics and not population genetics. Even with this interpretation it has frequently been accepted by comparative anatomists and naturalists as providing the simplest explanation for the occurrences which they can observe in the natural world. However, practically all experimentalists have rejected it. I need not summarize the well-known experiments which have failed to demonstrate an effect of environmental conditions on the hereditary qualities which are passed on from parent to offspring.

In quite recent years the situation has changed somewhat. We have now obtained abundant evidence of the induction of hereditary changes—in the form of gene mutations, chromosome aberrations, etc.—by external agents such as ionizing radiation and highly reactive chemicals. But these changes are non-directional; and induced mutagenesis as we normally encounter it in the laboratory does not provide any mechanism by which relatively normal environments could induce hereditary changes which would improve the adaptation of the offspring to the inducing conditions. Directional hereditary changes have, indeed, also been induced, but, so far, only in very simple systems such as bacteria, and by the use of highly specific inducing agents—for instance, the transforming principles. A more general mechanism of biological alteration, which does not depend on such exceptional inducing agents, is the induction of the synthesis of specific enzymes related to particular substrates. The changes produced in enzyme induction are for the most part not hereditarily transmissible, but it seems in principle not inconceivable that under suitable conditions actual gene mutations could be induced by some such mechanism.

Finally, one should notice some recent evidence which has been produced to support the hypothesis that variations of the normal environment may in some cases induce hereditarily transmissible changes. This evidence relates largely to plants, and most of it has emanated from Russia and is regarded with considerable scepticism in other countries, where attempts to repeat the experiments have been rather uniformly unsuccessful. Nevertheless, evidence of a not entirely dissimilar character has begun to appear in Western countries also—for instance, in the studies of Durrant[2] on the hereditary transmission of the effects of manurial treatment of flax, the work of Highkin [3] on the effects of alternating temperature on peas, and a few others. It is not clear in any of these cases that the hereditary effects produced, if any, are of a kind that improves the adaptation of the organism to the inducing conditions. The field of work is clearly one of great inherent interest, but it remains true that the vast majority of changes in the environment do not directly produce any hereditary modifications in the organisms subjected to them, and we are certainly very far from being able to provide a general explanation of evolutionary adaptations in terms of the type of effects which have just been mentioned.

The development of evolutionary theory in the last hundred years has in fact proceeded along quite other lines. Darwin's major contribution was, of course, the suggestion that evolution can be explained by the natural selection of random variations. Natural selection, which was at first considered as though it were a hypothesis that was in need of experimental or observational confirmation, turns out on closer inspection to be a tautology, a statement of an inevitable although previously unrecognized relation. It states that the fittest individuals in a population (defined as those which leave most offspring) will leave most offspring. Once the statement is made its truth is apparent. This fact in no way reduces the magnitude of Darwin's achievement; only after it was clearly formulated, could biologists realise the enormous power of the principle as a weapon of explanation. However, his theory required a second component— namely, a process by which random hereditary variation would be produced. This he was unable himself to provide, since the phenomena of biological heredity were in his day very little understood. With the rise of Mendelism, the lacuna was made good. Heredity depends on chromosomal genes, and these are found in fact to behave as the theory requires, altering occasionally at unpredictable times and in ways which produce a large, and, it is usually stated, 'random' variety of characters in the offspring bearing the altered genes. On these two foundations—natural selection operating on variation which arises from the random mutation of Mendelian genes—the present-day neo-Darwinist or 'synthetic' theory of evolution has been built up.

This theory has brought very great advances in our understanding of the genetic situation in populations as they exist in nature, of the ways these genetic systems may change, and of the differences between local races or between closely related species. The question discussed in this paper is the adequacy of its treatment of the major problem of the 'appearance of design', or biological adaptation. In dealing with this problem, neo-Mendelian theory relies essentially on the hypothesis that genes mutate at random, that is to say, if one waits long enough, an appropriate gene mutation will occur which will modify the phenotypic appearance of the organism in any conceivable way that may be required. It is pointed out that, however rare such a mutation may be, the mechanism of natural selection is eminently efficient at engendering states of high improbability, so that, from rare and entirely chance occurrences, an appearance of precisely calculated design may be produced.

This explanation is a very powerful one. It could, in fact, explain anything. And there is no denying that the processes which it invokes —random gene mutation and natural selection—actually take place. I should not dream of denying it—as far as it goes—but I wish to argue that it does not go far enough. It involves certain drastic simplifications which are liable to lead us to a false picture of how the evolutionary process works, whereas, if we take into account certain factors which have been omitted from the conventional picture, we shall not only be closer to the situation as it exists in nature but will find ourselves with a more convincing explanation of how the appearance of design comes about.

Let us consider some examples of the type of biological adaptedness which we are trying to understand. In many cases in which we speak of an animal as being adapted, the adaptation is comparatively trivial and its precise character is not critical. As an example, one may take the phenomenon of industrial melanism in Lepidoptera, which is one of the best-studied examples of natural selection in the field. In industrial areas of Great Britain several species of moths which a century ago most commonly appeared in fairly light-coloured forms have in recent years shown increasing numbers of dark melanic varieties; in several instances these have now become by far the most common type in regions contaminated by industrial fumes. A typical region where this replacement of light by dark forms has occurred has earned the nickname 'The Black Country'. It was natural to suppose that the blackening of the moth is connected with the darkening of the general vegetation by contamination with industrial smoke.

One of the earliest investigators to examine the subject in some detail was Heslop-Harrison[4]. He held somewhat Lamarckian views about the origin of the melanic form and, indeed, claimed at one time to show that it could be induced by feeding larvae on leaves which had been contaminated by various metallic salts. This claim has not found general acceptance in later years. However, Heslop-Harrison also made a further and more important contribution to the subject. He demonstrated that natural selection operates differentially on dark and light forms of the moth *Oporabia autumnata*, the melanics being favoured in regions with a dark background whereas the light forms were favoured in the presence of light-coloured vegetation. He was able to show this with particular clearness by studying a wood in which one section had been separated in about 1800 by the cutting of a wide gap, later grown up with heather. A considerable number of years later, in 1885, the southern section of the wood was planted with light-coloured birch trees, while in the northern portion nearly all the trees were dark pines. By 1907 the populations of moths in the two sections of the wood showed a quite different proportion of dark to light forms. In the birch part of the wood only 15 per cent were melanics; in the pine section about 96 per cent were dark. Moreover, Heslop-Harrison showed that in the dark pinewood section, where by far the majority of the moths were melanics, the majority of the wings found isolated on the ground —representing remnants left after the insect's body had been eaten by a predator—were actually light. Thus it was clear that the pale-coloured forms were at a disadvantage in the dark wood.

A similar situation has been re-examined in a much more thorough form quite recently by Kettlewell[5]. He found that the predation is in this case, at least to a large extent, carried out by birds, and he has been able to demonstrate very clearly the reality of the natural selective advantage enjoyed by an insect which blends reasonably well with its background. We have here, then, a well-studied example of an evolutionary process in which a species acquires a characteristic—namely, melanism—which can be considered to adapt it to its surroundings. However, in this example the adaptive character is of the very simplest kind. It is a mere darkening of the wings, and it does not seem at all likely that the precise pattern in which the blackening is laid down can have any great importance. The effective change probably involves nothing more elaborate than a markedly increased production of the melanic pigment. It is perhaps satisfying enough in such a case to attribute the appearance of the relevant new hereditary variation simply to a random gene mutation.

But the adaptations which have tempted man to think of design are rather more far-reaching. For instance, figure 1 shows the skeleton forelimbs of a gibbon and a pangolin. The former uses its arms for climbing in trees, the latter for digging in hard soil. The limb bones are very precisely modelled in relation to the functions they will carry out, and the difference between them involves something more than a simple over-all change comparable to the blackening of a moth's wing. A mere lengthening of the pangolin's arm would not turn it

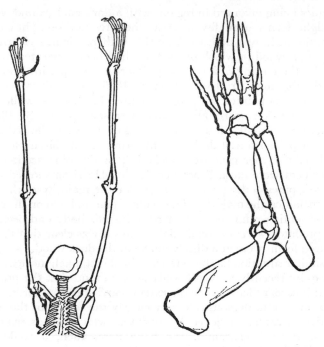

Figure 1. The skeleton forelimbs of a gibbon (left) and of a pangolin.

into that of a gibbon. We are dealing here with a precise set of carefully co-ordinated changes involving several different bones of the limb and the shoulder girdle. Now we know that during the lifetime of any single individual the use of the limb muscles in a particular way will increase the size and strength of those muscles, and if the operations take place early enough, they have some effects on the associated bony structures, these effects being of a co-ordinated kind, such as those which distinguish the two forelimbs illustrated. In conventional neo-Mendelian theory, these effects of the use of an organ, exerted during the organism's own lifetime, are dismissed as irrelevant to the evolutionary process. They are 'acquired characters' and are not genetically inherited. In so far as the individual exhibits them, his phenotype is likely to deceive us as to the characters which he will pass on to his offspring. The acquired characters act, the neo-Mendelian theory asserts, merely as genetic 'noise', in the information-theory sense of that term. We have to find the explanation for such evolutionary changes in random gene mutations, to whose occurrence the physiological processes which lead to the formation of adaptive ontogenetic changes are completely irrelevant.

However, that explanation leaves us with two major points on which we may feel some lack of satisfaction. One is that we have no specific explanation for the co-ordinated nature of the changes as they affect the different bones or other subunits in the system. Can we do no better than fall back on the very general explanation in terms of the efficiency of natural selection in engendering highly improbable states? Since we see similar co-ordinated changes being produced by physiological adaptation within a lifetime, this highly abstract principle seems a little inadequate.

Second, we are bound, both for practical reasons and on the basis of fundamental theory, to regard all forms, functions, and activities of an organism as the joint product of its hereditary constitution and its environmental circumstances; the exclusion of acquired characters from all part in the evolutionary process does less than justice to the incontrovertible fact that they exhibit some of the hereditary potentialities of the organism. All characters of all organisms are, after all, to some extent acquired characters, in the sense that the environment has played some role during their development. Similarly, all characters of all organisms are to some extent hereditary, in the sense that they are expressions of some of the potentialities with which the organism is endowed by its genetic constitution.

This point is one which has only recently forced itself firmly into the attention of geneticists, perhaps largely through current interest in characters to whose variation hereditary differences contribute only a small fraction, such as the milk yield of cattle. It is still not always kept in mind in all contexts in which it is relevant, and in the earlier days of genetics it was very frequently ignored. To take a relevant example from the very early years of this century, Baldwin and Lloyd Morgan pointed out that a capacity for carrying out adaptive changes during their lifetimes might enable organisms to survive in environments in which they would otherwise be inviable and that they could in this way exist until a suitable hereditary variation occurred which could be seized upon by natural selection and enable a genuine evolutionary adaptation to take place. They did not point out that there was any hereditary variation in the capacity for forming such ontogenetic adaptations. Mayr[6] actually describes their view as 'the hypothesis that a non-genetic plasticity of the phenotype facilitates reconstruction of the genotype'. But a plasticity of the phenotype *cannot* be 'non-genetic'; it *must* have a genetic basis, since it must be an expression of genetically transmitted potentialities. It is conceivable, of course, that in any given population there will be no genetic variation in the determinants of this plasticity, but our experience of natural populations shows that this is a very unlikely state of affairs. When wild populations have been investigated, they have, I think, always exhibited some genetic variation in respect of any character that has been studied.

There is actually no need to rely on purely *a priori* arguments in this respect. Experiments have recently been made in which *Drosophila* populations were searched for the presence of genetic variation in the capacity to respond by ontogenetic alterations to the stimuli produced by various abnormal environments, and the effectiveness of selection of this genetic variation was studied. Both the characters involved and the environmental stimuli applied were of rather diverse kinds in the different experiments in the series, but in all cases genetic variation in capacity to response, utilizable by selection, was revealed.

Perhaps the simplest of these experiments was actually the last to be performed[7]. It attempted to bring about the adaptation of a population of *Drosophila* to a high concentration of sodium chloride in the medium in which the larvae live. The larvae possess anal papillae on either side of the anus, and these are known to play some part in regulating the osmotic pressure of the body fluids[8]. The size of the papillae can be measured most accurately just after pupation when the hardening of the puparial skin prevents distortion by the muscular movements of the body. Three stocks were employed. One was a wild-type 'Oregon K'; the other two, sp^2 bs^2 and al b c sp^2, each contained the gene speck in which the anal area is pigmented in the pupa, making it somewhat easier to see the anal papillae (figure 2). From the Oregon wild-type stock two selected

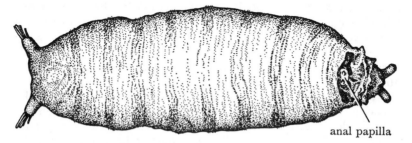

anal papilla

Figure 2. A pupa from a stock containing sp^2, showing the pigmented anal area.

lines were set up, known as 'Oregon L' and 'Oregon E', while for each of the other two stocks one selected line was maintained. These selected lines were carried on by growing the larvae of each genera-ation on normal *Drosophila* medium to which various concentrations of sodium chloride had been added, the concentrations being adjusted so that only 20–30 per cent of the eggs laid on the medium survived to the adult condition. No artificial selection was made, the selection pressure being entirely the natural selection exerted by the stringent medium.

After 21 generations of selection in this way, the survival of the various strains on different concentrations of salt was tested and the mean size of the anal papillae estimated by measurements on 20 individuals from each culture at each concentration of salt (figures 3 and 4). The selected stocks became somewhat more tolerant of high salt concentration, though the difference was not very great. However, there is no doubt that some genetic variability exists in the capacity of the animals to adapt themselves to the environmental stress and that this genetic variability has been utilizable by the natural selection employed.

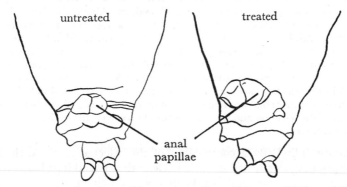

Figure 3. Outline drawings of the anal papillae of 'Oregon K' *Drosophila* larvae representing extreme variants in size, the larger from a selected strain grown on a medium with seven per cent salt added, the smaller from an unselected strain grown on normal medium with no added salt.

A number of further deductions can be made from figure 4. In the first place, it is clear that the size of the anal papillae tends to increase with increasing concentration of salt in the medium, although the effect is rather slight until the concentration reaches a high level. For any one stock the curve relating size of papillae to the salt concen-tration gives a picture of a physiological function which we might call

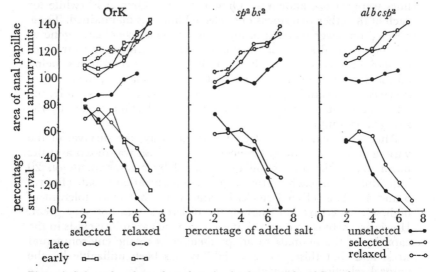

Figure 4. Selected and unselected strains in three stocks of *D. melanogaster*, in relation to the salt content of the larval medium. Above, the size of the anal papillae at various concentrations, in units derived from micrometer measurements. Below, the percentage of adults appearing from a given number of eggs. For the wild-type stock, two selected strains were prepared, one selected also for early emergence and the other for late. The papillae of the selected stocks were measured both in larvae derived from parents grown in the selection-medium (seven per cent added salt) and in 'relaxed' lines in which there had been one generation on normal medium between the end of the selection and the setting-out of larvae on the various concentrations.

its 'adaptability'. By a comparison of the selected stocks with the corresponding unselected ones, it is clear that two things happen to these curves of adaptibility. In the first place, their steepness increases and to some extent their general shape changes; that is to say, the selection favours, as might be expected, those genotypes which endow the individual with a relatively high capacity to carry out an onto-genetic adaptation to the stress of high salt content. Second, the general level of the curves is raised. One might refer to this general level as the 'level of adaptation to high salt content'; we can say, therefore, that the level of adaptation has been increased. A third, and perhaps most important point, is that the anal papillae in the selected races remain larger, even at low salt content, than the papillae of the unselected strains at the same concentration. The adaptation to high salt content which has been produced by 21 generations of selection is not immediately reversible by one or two generations of the normal medium. In the botanical terminology employed by Turesson[9], the ecotype which has been produced in relation to high salt concentration is to some extent an ecogenotype. The character of the adapted strain depends, of course, on its genotype, as all characters of all strains do, but the point to notice is that the genetic difference between the selected and unselected strains is expressed also in the normal low salt medium. We have obtained a result which is effectively the same as would have resulted from the direct inheritance of acquired characters but which has been produced, not by the mechanisms which are usually thought of in connection with Lamarck's hypothesis, but by a population-genetical mechanism which involves selection.

The failure of the selected strains when grown in normal medium to revert completely to the condition of the unselected strains must depend on a certain inflexibility of their developmental processes. Although their adaptibility becomes higher, as we have seen, it is not large enough to allow the anal organs to regress completely on the low salt medium. Such lack of flexibility in the developmental system has been referred to as 'canalization'[10,11].

It is sometimes useful to discuss development in terms of a diagram in which the course of normal development is represented as the bottom of a valley, the sides of which symbolize the opposition that the system presents to any stresses which attempt to deflect development from its normal course. A cross-section of the valley represents, in fact, the curve that we have defined as the adaptability of the system, with the minor modification that the scale on which the stress is represented is reversed as between the two sides of the valley, so that it measures divergencies from the normal—below it on one side or above it on the other. The surface which in this way symbolizes the developmental potentialities of the genotype has been called the 'epigenetic landscape' [see figure 4, p.259].

Figure 5 shows an interpretation in terms of this model of the assimilation of a character in the development of which no threshold phenomena are involved, as is the case with the anal papillae. This diagrammatic form is particularly appropriate in situations involving developmental thresholds, which have been studied in some other *Drosophila* experiments. In these, quite abnormal environmental stresses were applied to the developing system, and artificial selection was made for certain categories of response. Although both the stresses and the selection were artificial, these experiments reveal a type of process which might go on in natural populations under the influence of natural stresses and natural selection.

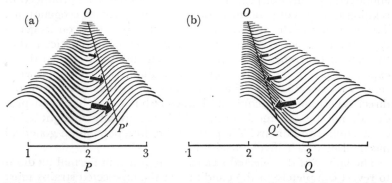

Figure 5. Assimilation of a character not involving a threshold. In the initial situation (a), an organism, which contains an organ which normally develops along a canalized pathway to end state *P* at 2, has moved into an unusual environment which diverts development to *P'*. After some generations of selection in this environment, the developmental pathway has been remodelled (b) so as to lead to a new end-state *Q*, located at 3, which is taken as the optimum for this environment. When organisms with this selected genotype are placed back in the original environment, their development becomes diverted to *Q'*; that is, it does not revert all the way back to the value 2 which it initially had. Some of the adaptive response has been genetically assimilated.

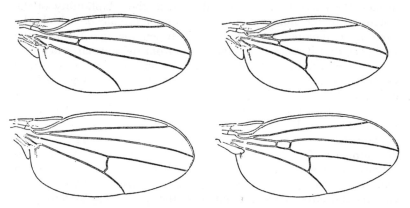

Figure 6. Some types of venation phenocopies induced by a heat shock to the 18-hr pupa in *D. melanogaster* [13].

In the first experiment [12, 13], a heat shock was applied to pupae of an age which was known to be suitable for producing a number of phenocopies affecting the cross-veins. In point of fact, several different phenocopies appeared, involving absence of one or another of the cross-veins or in some cases increases in venation (figure 6). If selection was exercised for any specific one of these types of phenocopy, strains could be rapidly built up which responded to the standard stress by a high frequency of this particular developmental abnormality. Moreover, after fairly intensive selection it was possible to produce strains in which the particular modification which had been selected for appeared in high frequency even in the absence of the stress. We had again carried out the process, which I have called 'genetic assimilation', by which selection produces genotypes which modify development in the same manner as did the original environmental stress.

An attempt was also made to produce the genetic assimilation of a very remarkable phenotypic modification which, if it appeared in nature, would probably be considered of macro-evolutionary importance [14, 15]. If the eggs of a normal wild-type *Drosophila* stock are treated with ether vapour soon after laying, a certain proportion of them develop a bithorax phenotype (figure 7; cf. [16]). If one exerts artificial selection for the capacity to respond to this particular environmental stress, one can increase the frequency of the response—or, by selecting against it, decrease it. Again, after something over 20 generations of selection, it was possible to produce

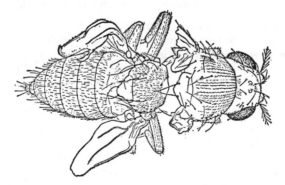

Figure 7. The bithorax phenotype as developed after ether treatment of the egg. The main wings have been removed to show more clearly the transformed metathorax. The individual depicted is actually from the 'assimilated' stock and developed *without* ether treatment [14].

an assimilated bithorax stock in which the phenotype is developed
in high frequency even in the absence of any ether-vapour treatment
(figure 8).

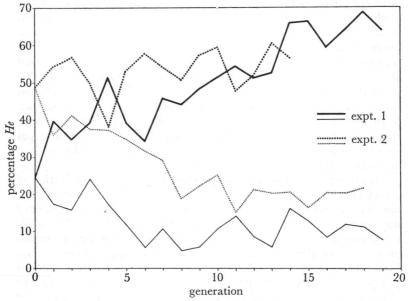

Figure 8. The progress of selection for or against bithorax-like response to
ether treatment. Two experiments are shown, starting from two wild-type
populations which reacted with rather different frequencies [14].

It seems profitable to discuss these last two experiments in terms
of the canalization model mentioned above [17]. We can picture the
development of the cross-vein region (or of the thorax) proceeding
under normal circumstances along a certain valley leading to the
normal adult condition (figure 9). The slope of the sides of the valley
towards the bottom means that the system is to some extent resistant
to stresses which might tend to produce an abnormal end result. The
fact that the system responds by phenocopy formation to certain
stresses applied at definite times can be represented by drawing a
side valley, reached over a col, at that time in development. The
particular configuration of the surface drawn at the top left of figure
9 represents the developmental potentialities of one specific genotype.
In any large population the genetic variation in the frequency with
which the response occurs will correspond to variations in the height
of the col above the main valley floor. Similarly, variation in the type

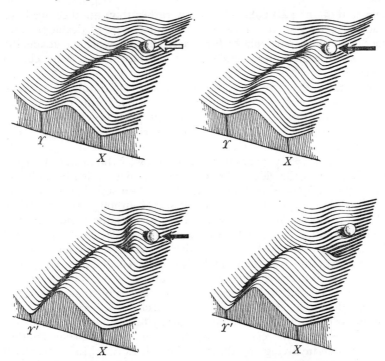

Figure 9. Modification of the epigenetic landscape by selection. The upper left drawing shows the situation in the unselected foundation stock; a developmental modification Y will occur only if an environmental stress (white arrow) forces the developing system to cross a threshold or col. The upper right figure shows the Baldwin-Lloyd Morgan hypotheses—that a new gene mutation (black arrow) appears which substitutes for the environmental stress, everything else remaining unaltered. The two lower figures show stages in the selection of genotypes in which the threshold is lowered (requiring only a 'small' gene mutation or, eventually, a single specifiable mutation) and the course of the developmental modification is made more definite and directed to the optimal end-result, Y' [17].

of phenocopy produced (an absence of the posterior or the anterior cross-vein, etc.) will be represented by variations in the course of the side valley. Selection, as we have seen, has been able to utilize both types of variation. In the assimilated stocks we have selected and combined low-col genes until we have reached a condition in which the col is non-existent and the floor of the upper part of the main valley leads off into what was originally the side branch. In the selection of one particular phenocopy rather than another, we have selected genotypes in which one particular type of developmental modification is particularly favoured; that is, we have made the course of the side valley more definite and have led it to our chosen end point.

We may ask ourselves where this genetic variability has come from. Was it perhaps created during the course of the experiment? There is rather good reason to believe that this was definitely not the case in the work involving the cross-veinless phenocopies. For instance, cross-veinless types occur spontaneously in some wild stocks; that is to say, genes which tend to lower the height of the col which defends the side valley are present in sufficiently high frequency for an occasional individual to contain sufficient of them to abolish the col even before selection starts, although the frequency of such combinations is so low that natural selection would scarcely be able to utilize them in the absence of the reinforcement produced by the environmental stress. Again, when selection for environmental response was made in highly inbred stocks, no effect was produced, indicating that new mutations were not occurring frequently enough to be effective (figure 10). This experiment also shows definitely that no direct Lamarckian inheritance of the acquired character is occurring in this system.

In the bithorax experiments, however, and also in another experiment involving the dumpy-wing phenotype [13], there is a strong suggestion that genes acting in the direction of selection turned up during the course of the experiments. Since the experiments involved some hundreds of thousands of flies, the occurrence of such mutations is not so unexpected that we have to attribute it to the environmental stress itself. The mutations presumably occurred in the normal manner, that is, 'at random'. But although the change in the chromosomal nucleoprotein may well have been quite undirected, the phenotypic effects of the genes were certainly influenced by the selection which had been practiced on the stock. The new 'bithorax-like' gene has a strong tendency to produce bithorax phenotypes (actually by a maternal effect, but that is irrelevant to the present discussion) when in the genetic background of the selected stock but only a very weak tendency to do so in a normal, unselected, wild-type background. A similar consideration applies to the 'dumpy-like' gene. Selection, if you like, by reducing the height of the col and making the side valley more definite, has produced genotypes whose developmental potentialities are such that the course of development is easily diverted to the production of the particular adult condition that has been selected for. We have, as it were, set the developmental machine on a hair trigger. Quite a number of gene mutations, which are random at the level of nucleo-protein structure, are likely to produce this preset phenotype, and are therefore by no means random in their developmental effects.

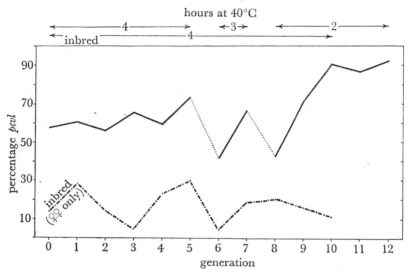

Figure 10. Selection for the frequency of formation of a broken posterior cross-vein in response to a temperature shock. Above, a wild-type stock, subjected for five generations to a 4-hour treatment, but later to treatments of only 3 and 2 hours; below, an inbred stock, subjected to 4-hour treatments throughout the experiment [13].

The particular importance of this conclusion concerns the evolutionary origin of *co-ordinated* effects on the subunits of a structure, of the kind which were illustrated in the forelimbs of the gibbon and pangolin. The adult form of any animal is the result of the interaction between its genotype and the environmental stimuli and stresses to which the developing system has been subjected. If one thinks of the stresses produced by life dependent on digging for food, it is clear that the stimuli may be very complex. When only a single stress is involved and the response of the developing system shows a certain approximation to an all-or-none character, as in the temperature-shock or ether-treatment experiments, one can represent the system by a diagram involving a single col or even a sharply defined thres-hold. When one considers the more complex stresses which arise in real life, such a representation becomes more difficult and also more artificial. But the essential point is that the complex stresses give rise to developmental responses which are co-ordinated. In a wild population these responses are of adaptive value, natural selection will occur and will increase not only the intensity of the response but also its co-ordination. It will build up genotypes whose developmental potentialities include a high capacity for producing a well organized and harmonious adaptive phenotype. This capacity may then be released by quite a variety of random changes in the nucleoproteins of the chromosomes.

In this way, by taking into account the possibility of selection for both capacity to respond and type of response to environmental stresses, we can once again find justification for attributing the 'appearance of design', or co-ordinated adaptations, to the epigenetic processes which we know to have co-ordinated effects; and we can reduce our dependence on the abstract principle that natural selection can engender states of high improbability. We have, in fact, found evidence for the existence of a 'feedback' between the conditions of the environment and the phenotypic effects of gene mutations. The 'feedback' circuit is the simple one, as follows: (a) environmental stresses produce developmental modifications; (b) the same stresses produce a natural selective pressure which tends to accumulate genotypes which respond to the stresses with co-ordinated adaptive modifications from the unstressed course of development; (c) genes newly arising by mutation will operate in an epigenetic system in which the production of such co-ordinated adaptive modifications has been made easy [17].

Before concluding, I should like to return to the earlier point that the stresses to which an animal will be subjected depend at least in part on its own behaviour. Nearly all animals live in surroundings which offer them a much greater variety of habitats than they are willing to occupy. Naturalists, of course, are very familiar with the fact that closely related species often show markedly different preferences for particular types of habitat; even within species, different races may exhibit relatively specific patterns of behaviour. These obviously play a considerable evolutionary role in connection with reproductive isolation (e.g. on *Drosophila*, [18, 19, 20]). They may also affect more general choice of living conditions (e.g. [21]), but this field is still very incompletely explored. For instance, it is clear that cryptic coloration is of very little use to an animal unless its behaviour is such that it makes use of the possibilities of concealment which are offered to it, but we know little about the genetic

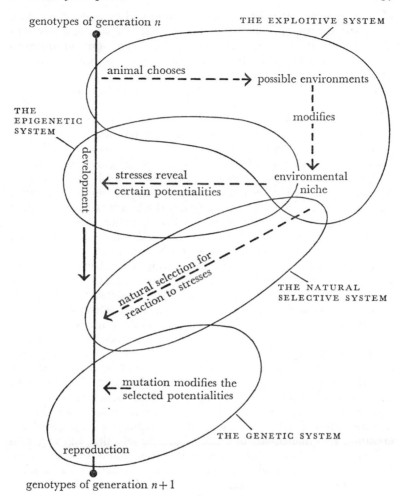

Figure 11. The logical structure of the evolutionary system. Changes in gene frequency between successive generations involve the operation of four subsystems: the exploitive, the epigenetic, the natural selective, and the genetic [7].

correlations, if any, between the production of cryptic coloration and the appropriate types of behaviour, although Kettlewell[22] has shown that melanic moths do in fact tend to settle on the darker areas of trees more frequently than would be expected by chance. It

is clear, however, that here again selection will be operating not on
the isolated components—behaviour on the one side and developmental and physiological response on the other—but on an inter-
locking system in which behaviour and other aspects of function
mutually influence one another.

The result of this discussion is to suggest that we have perhaps been
tempted to oversimplify our account of the mechanism by which
evolution is brought about. This mechanism—the evolutionary
system, as it may be called—has often been envisaged as consisting
of no more than a set of genotypes which are influenced, on the one
hand, by a completely independent and random process of mutation
and, on the other, by processes of natural selection which again are in
no way determined by the nature of the genotypes submitted to them.
Perhaps such a simplification was justified when it was a question of
establishing the relevance of Mendelian genetics to evolutionary
theory, but it can only lead to an impoverishment of our ideas if we
are not willing to go further, now that it has served its turn.

In point of fact, it would seem that we must consider the evolutionary system to involve at least four major subsystems (figure 11). One
is the 'genetic system', the whole chromosomal-genic mechanism of
hereditary transmission; the second is natural selection; a third,
which might be called the 'exploitive system', comprises the set of
processes by which animals choose and often modify one particular
habitat out of the range of environmental possibilities open to them;
and the fourth is the 'epigenetic system'—that is, the sequence of
causal processes which bring about the development of the fertilized
zygote into the adult capable of reproduction. These four component
systems are not isolated entities, each sufficient in its own right and
merely colliding with one another when impinging on the evolving
creature. It is inadequate to think of natural selection and variation
as being no more essentially connected with one another than would
be a heap of pebbles and the gravel-sorter onto which it is thrown. On
the contrary, we have to think in terms of circular and not merely
unidirectional causal sequences. At any particular moment in the
evolutionary history of an organism, the state of each of the four main
subsystems has been partially determined by the action of each of the
other subsystems. The intensity of natural selective forces is dependent on the condition of the exploitive system, on the flexibilities and
stabilities which have been built into the epigenetic system, and so
on.

Very much remains to be done in working out the theory of evolution from this more inclusive point of view. But one general point is

already clear. We can now see that the system by which evolution is brought about has itself some degree of organization, in the sense that its subsystems are mutually interacting, and, in fact, mutually interdependent. In the recent past we have been working with a theory in which the obvious organization of the living world has to be engendered *ab initio* out of non-organized basic components—'random' mutation, on the one hand, and an essentially unconnected natural selection on the other. We had to rely on a Maxwell demon, and persuade ourselves not merely that natural selection could show some of the properties of such a useful *deus ex machina* but that it had them so fully developed that we needed nothing further. This was a rather uncomfortable position, and we can now escape from it.   [1959]

GENETIC ASSIMILATION

'Genetic assimilation' is a name which has been proposed[1] for a process by which characters which were originally 'acquired characters', in the conventional sense, may become converted, by a process of selection acting for several or many generations on the population concerned, into 'inherited characters'.

Since the earliest days of evolutionary thought, the problem of 'the inheritance of acquired characters' has been a central subject of debate. With the rise of Mendelian genetics in this century, it has often been considered that the problem has finally been resolved, in the sense that acquired characters are not inherited, and that therefore, the fact that characters may be acquired has no direct influence on the course of evolutionary change. This view has been rather generally accepted among geneticists, but has still found some opponents among naturalists, some of whom have felt that the phenomena of adaptation seen in the organic world force them to conclude that the acquirement of characters must have evolutionary consequences, presumably through some unknown process by which these characters are in fact inherited by later generations. The phenomena revealed by the experiments on genetic assimilation cast no doubt on the thesis, generally accepted by geneticists, that acquired characters are not inherited (except in very special circumstances), but lead to the conclusion that there is no justification for arguing from this that they have no effect on the course of evolution. On the contrary, it becomes apparent that the conventional and accepted facts and theories of genetics provide a mechanism by which 'acquired characters' must exert some influence

—and probably a rather important one—on the direction in which evolutionary change proceeds.

This article, therefore, will not be concerned with revealing new genetic principles, but with describing a type of process which, although it is new in the sense that it had not been contemplated until a few years ago, derives its flavour of novelty or unexpectedness only from the fact that normal genetic ideas have very often been misinterpreted or incompletely thought out in this context. In order to clarify these misinterpretations, it will be necessary to pay a good deal of attention to the definitions and content of various genetic notions. The general plan of the article will be, first, to provide some of the necessary definitions, then to illustrate the application of these by reference to the experiments which have been conducted on genetic assimilation, and finally to consider some of the wider repercussions of the ideas which have been developed.

DEFINITIONS

Acquired and inherited characters

The notions of 'acquired' and 'inherited' characters, as they are employed in evolutionary theory, are not entirely straightforward. As many authors, e.g. Goodrich[2], have remarked, all characters of all organisms are to some extent 'inherited', in the sense that they can only be developed if the organism contains the hereditary potentialities for developing them, and are also to some extent 'acquired', since all development involves some participation of the environment (cf.[3, 4]). Such statements have often been dismissed as quibbling truisms, but the train of thought which led to the realization that genetic assimilation may occur arose from taking seriously the fact that 'characters' are produced by the joint action of genotype and environment.

The definition of acquired and inherited characters, in the sense in which those phrases are normally used, may be put as follows. If we have reason to believe that an organism, if reared in environment E, would exhibit the phenotype P, while if reared in environment E' its phenotype would be P', then those features in which P' differs from P are considered as acquired characters, and those features in which P' resembles P are considered as inherited characters, with reference to this particular change of environment.

Clearly, one of the difficulties in the application in practice of this definition turns on the phrase 'have reason to believe'. We cannot rear one and the same organism in two different environments. The best we can do is to place in the two environments two organisms, or populations of organisms, which are thought to have effectively similar genotypes. It is only when the two test individuals or groups can be taken from one inbred or clonally propagated population that we can be confident of their genetic identity. When we are dealing with individuals from a population which is crossbred, and therefore somewhat heterogeneous genetically, we must expect to find that the difference between P and P' will not be exactly the same for all pairs tested. That is to say, there will be some variation in the degree to which any given character is an acquired or an inherited one, according to the particular test-pair used.

Genetic assimilation

The notion of genetic assimilation involves both a phenomenon, and a mechanism by which this phenomenon is brought about.

The phenomenon may be described as the conversion of an acquired character into an inherited one; or better, as a shift (towards a greater importance of heredity) in the degree to which the character is acquired or inherited. Consider a population living in environment E and exhibiting phenotype P. Let a subpopulation be placed in environment E', where it exhibits a phenotype P'. Then P–P' is an acquired character in this subpopulation. Now, after the subpopulation has continued in E' for a greater or lesser number of generations, let it be returned to environment E; and suppose that it there exhibits phenotype P''. Then the degree to which P'' resembles P' is a measure of the extent to which the original acquired character P–P' has been converted into an inherited character. The name 'genetic assimilation' is given to such processes of conversion when they are brought about by selection acting on the genotypes in the subpopulation which was transferred from E to E'.

As we shall see, other mechanisms of conversion have been suggested (e.g. the 'Baldwin Effect'), but there is considerable doubt whether they could possibly occur.

EXPERIMENTAL EVIDENCE FOR GENETIC ASSIMILATION

After these abstract definitions, it will be as well to consider next the experimental evidence for the reality of the process of genetic assimilation. The process was first envisaged as a theoretical possibility nearly twenty years ago[5] and reference was made to it in a number of later general discussions of evolutionary problems [6–13].

Characters involving a developmental threshold

The first experimental demonstration of the process was not published till some years later [14, 1]. These experiments started with a wild-type population of *Drosophila melanogaster* which, when reared in the normal laboratory culture environment, exhibited the well-known normal phenotype. A subpopulation was placed in another environment E', which differed from the normal in that a high-temperature shock (four hours at 40°C) was given at 17–23 hours after puparium formation. It was found that some of the sub-population treated in this way exhibited an abnormal phenotype P', which took the form of the breaking of the posterior crossvein on the wings. This 'acquired character' was exhibited by only a proportion of the subpopulation. When selection was applied within the sub-population, by taking only those individuals with broken crossveins for further breeding, the frequency of the crossvein defect increased from generation to generation; similarly, with selection against the appearance of the abnormal phenotype, it decreased in frequency. Thus the subpopulation must have contained genetic variation concerned with the development of the crossveinless phenotype in the environment which included the temperature shock.

After a few generations of selection, it was discovered that the most effective period for the administration of the temperature treatment was between 21 and 23 hours after puparium formation. Using this treatment as the abnormal environment E', selection proceeded rapidly, so that, after about a dozen further generations of upward selection, well over 90% of the individuals developed with broken crossveins following temperature treatment. In each generation a fair number of flies were examined which had developed from untreated pupae, that is to say, which had been kept all the time in the original normal environment. No crossveinless individuals were found among these until generation 14, when a few isolated cases appeared in the strains being subjected to upward selection. In generation 16 of upward selection there were 1–2% crossveinless individuals among the flies developing in the normal environment without heat shock, and from these a number of pair matings were set up. By selection over a few generations among the progeny of these pair matings, a number of different strains were derived, some of which produced high frequencies of crossveinless individuals when reared in normal environments lacking any heat shock treatment. In some of the 'best' of these lines, in fact, the frequency of crossveinless individuals was 100% at 18°, though somewhat lower at 25°.

In these lines, therefore, the character of crossveinless, which had originally been an acquired character only exhibited (at any rate, at noticeable frequencies) in the abnormal environment provided by the heat shock, had been converted into an inherited character, exhibited in the normal environment in the absence of the heat shock. It was clear that this conversion had been brought about by the selection applied, and the process was therefore considered to be one of genetic assimilation, as defined above.

In these first experiments, the relatively complete assimilation finally achieved was obtained by selection operated in two different phases. In the first phase, selection was exerted on the subpopulation submitted to the normal environment of the temperature shock. After a certain modest degree of assimilation had been achieved in this way, the second phase of selection was carried out, in the normal environment, on the strains in which this mild assimilation was exhibited. This was done because a greater intensity of selection could be achieved by carrying out the second phase in this way, so that progress toward complete assimilation would be more rapid.

A series of very similar experiments were then made by Bateman [15–18]. She also employed a temperature treatment of the pupae as the abnormal environment. A number of different developmental abnormalities were chosen as the acquired characters to be favoured by selection. One of these was the breakage or absence of the posterior crossvein, which had been studied by Waddington. Others were (a) the absence of the anterior crossvein, (b) the appearance of an extra crossvein in the first posterior cell, (c) the presence of an extra crossvein in the submarginal cell, (d) the appearance of a distorted wing resembling dumpy. Similar experiments involving temperature treatment of the pupae and selection for absence of posterior crossvein were also made by Milkman[19]. Waddington [20, 21] used a very different abnormal environment (ether treatment of newly laid eggs) to elicit a different type of developmental abnormality—the conversion of the metathorax into a structure resembling the mesothorax, as in the well-known bithorax phenotype.

In all these cases, essentially similar results were obtained. Selection for a particular acquired character within a population subjected to an abnormal environment increased the frequency with which that character appeared in later generations. After a varied number of generations of such selection, individuals exhibiting the phenotype began to appear in samples of the selected stock which had not been subjected to the unusual environment, and from these individuals strains exhibiting high degrees of genetic assimilation could rapidly be established. Even without resorting to this second stage of selection, Bateman was able, by selection only within the abnormal environment, to raise the percentage of individuals which showed the character concerned, when grown in the normal environment, to the level of 12–24%, and in one case as much as 46%.

This provides strong grounds for believing that the use of second phase selection is not essential for the success of genetic assimilation, but is, as had previously been thought, merely a way of speeding it up.

Characters not involving a threshold

In all the above experiments, the acquired character was one whose development involves something in the nature of a threshold; that is to say, in many individuals in the treated populations it is not exhibited at all—the crossvein remains unbroken or the metathorax shows no signs of being modified into a mesothorax. The genetic assimilation of a character not involving a theshold has recently been investigated by Waddington[22]. This case is further interesting for two other reasons. Firstly, the selection applied was natural selection, brought about by the action of the abnormal environment itself and not by artificial intervention of the experimenter; and, secondly, the character produced is one which appears to be of adaptive value to the organisms when living in the abnormal environment.

The abnormal environment in this case was that produced by adding salt (sodium chloride) to normal *Drosophila* food. This was added in quantities (about 6% to begin with) sufficient to cause considerable mortality to the larvae reared on such media. Strains were carried on with no artificial selection, but simply by breeding from those individuals which had survived this stringent natural selection. The acquired character which was investigated was the size of the anal papillae. These were known to be concerned in the regulation of the osmotic pressure of the hemolymph, although their exact mode of operation is still obscure. It was found, in practice, that *Drosophila* larvae reared in media containing added salt have slightly larger anal papillae (corrected for variations in total body size) than those reared on normal media.

The degree to which the acquired character had become assimilated after 21 generations of natural selection was tested by allowing eggs from the selected strains to develop on media containing varied proportions of salt and measuring the size of their anal papillae. In all three strains tested the results were essentially similar. In the first place, the selection had resulted in an improvement in the adaptation of the strains to high salt media, as shown by the fact that in such conditions a higher proportion of the eggs from the selected strains developed through to adults than from the unselected strains. Further, in all three strains, the acquired character—enlarged anal papillae—had become to some extent genetically assimilated, since when the selected strains were grown in media with low salt content these organs were larger than in the unselected strains grown in the same media. The selection had also led to an improvement in the capacity of the strains to react to increased salt content of the medium by the development of larger papillae. This is shown by the fact that the curve relating size of papillae to salt content of medium rises more steeply in the selected strains than in the unselected.

The nature of the physiological change involved in this adaptation is little understood. Croghan and Lockwood[23] showed that the hemolymph of larvae of the selected strain, grown on food containing 7% added salt, is markedly hypotonic to the medium, the osmotic pressure not being much greater than that of unselected larvae on normal food; but since they did not study the blood of unselected larvae on salted food, the significance of their observations is uncertain.

All these experiments demonstrate that if selection takes place for the occurrence of a character acquired in a particular abnormal environment, the resulting selected strains are liable to exhibit that character even when transferred back to the normal environment. That is to say, the process which has been defined as genetic assimilation really occurs. In so far as this is true, the appearance of acquired characters which are of value to an organism in terms of natural selection will have evolutionary consequences. Natural selection for such characters will lead to the appearance of populations in which the character is an inherited one and will be developed even in environments other than that which originally provoked it and in which it is of adaptive value. We have, therefore, experimental justification for using the notion of genetic assimilation to explain all those evolutionary phenomena which people in the past have been tempted to attribute to the inheritance of acquired characters in the Lamarckian sense.

THE GENETIC MECHANISMS OF ASSIMILATION
Evidence that genetic assimilation depends on selection
Since selection, either artificial or natural, was applied during the production of the strains exhibiting genetic assimilation, it is natural to hold it responsible for the occurrence of the phenomenon. The point was, however, specifically tested by Bateman in experiments in which an attempt was made to bring about genetic assimilation in inbred strains. In two such strains, in which genetic variability must have been very small or negligible, selection for the acquired character crossveinless was completely without effect and no trace of genetic assimilation occurred.

Bateman[18] also investigated the results of relaxing selection at various stages during the process of genetic assimilation. It was found that when selection was stopped in a strain in which, say, 60 or 80% of the individuals exhibited the acquired phenotype in the normal environment, then over the next few generations the frequency of occurrence of this phenotype declined, eventually settling down, after about a dozen generations, to some level well above that of the unselected stock but far from 100%. However, in strains in which assimilation was complete, the abnormal phenotype appearing in 100% of individuals raised in normal environment, the relaxation of selection was without much effect. It may be concluded from this that the assimilation is due to gradual increase in frequency of appropriate genes, which eventually, in the completely assimilated stocks, reach 100%.

Analysis of the genotypes of assimilated strains

Genotypes of assimilated strains have been analysed by means of crosses and backcrosses between them, or between them and wild types or downward selected strains, and also by the use of chromosome markers[1, 17, 18, 20, 21]. All the evidence indicates that the differences between the assimilated strains and their foundation stocks always involve all the chromosomes. For instance, in his original experiments, Waddington[1] showed that in his assimilated crossveinless stock, the second and third autosomes and the X-chromosome all had a tendency to produce a crossveinless phenotype; the third chromosome had perhaps the strongest effect. The crossveinless-producing effect of fragments of the X-chromosome was also demonstrated for parts of the chromosome not containing the well-known sex-linked crossveinless locus. In crosses between the assimilated crossveinless stock and strains carrying the sex-linked *cv* factor a heterozygous female showed evidence of a summation of the effects of a single dose of *cv* and of an X-chromosome derived from the assimilated stock, but it did not appear that this chromosome contained any simple recessive allele of the *cv* locus. Waddington found further evidence of the participation of many genes in determining the character of the assimilated stock in the fact that the degree of dominance exhibited by the stock differed markedly in different crosses to various other stocks which had been derived during the process of selection. In crosses with downward selected stocks the crossveinless effect of the assimilated stock behaved almost as a complete recessive. On the other hand, when the assimilated stock was crossed to certain lines isolated during the process of assimilation, but themselves showing very low numbers of crossveinless individuals in normal environments, the assimilated genotype acted almost as a dominant.

Bateman [17, 18] also analysed the genetic constitution of several assimilated stocks, and again found that, in general, all the major chromosomes had some effect, although the intensity of effect varied from chromosome to chromosome. We may consider first the result of the analysis of stocks in which various venation phenotypes had been assimilated. Bateman analysed three stocks characterized by absence of the posterior crossvein, one with absence of the anterior crossvein and two with extra crossveins, either in the submarginal or the first posterior cell. In all three in which the posterior crossvein was absent, the third chromosome had the strongest effect, although in one of them the X-chromosome was also rather strongly effective. In the stock with absent anterior crossvein, the second and third chromosomes were of more or less equal, moderately strong, effect, while in the strains with extra crossveins, the second chromosome was the most important in both cases. As Bateman pointed out, a considerable number of loci which produce breakages of crossveins are known on chromosome 3, while on chromosome 2 there are quite a large number of genes which tend to produce extra veins. It still remains very obscure why genes producing these two types of phenotypic effect should be, as it were, sorted out into different chromosomes. Crosses were made between the assimilated crossveinless stocks and a number of laboratory stocks containing third chromosome factors such as *cv-c*, *cv-d*, and *det*. In some cases a fairly high percentage of crossveinless flies appeared in the F_1, but this could not be interpreted unambiguously to indicate that the assimilated stocks contained allelomorphs of the loci concerned, since similar results would be expected if the condition in the assimilated stocks had a multi-factorial basis.

The situation was rather different in the stock in which Bateman had assimilated the dumpy phenotype. In this case the abnormal environment was a treatment at 40°C for a period of 42 hours at 16–18 hours after puparium formation. In the early generations the incidence of the dumpy phenotype was erratic, but after a time the incidence rose considerably and by generation 30 of selection the phenotype occurred in about 90% of the treated flies after only two hours treatment. The beginning of genetic assimilation, i.e. the appearance of the phenotype in untreated individuals, did not occur till generation 25, but from these flies a fully assimilated dumpy stock was eventually derived. The genetic analysis of this stock revealed that it contained an allele of the dumpy locus extremely similar to, and perhaps identical with, the allele known as dp^{TP}. When this allele was removed from the assimilated stock, the remainder of the genotype was not able to produce any dumpy phenotypes in the normal environment. It could be shown that the assimilated stock contained other factors influencing the dumpy phenotype, since selection for severity of expression was effective, but, in this case, the assimilation depended on the presence of the particular relatively powerful dp^{TP} allele, and did not occur in its absence.

Genetic analysis of the various stocks derived during the assimilation of the bithorax phenotype also revealed some points of interest [21, 22]. In the attempt to assimilate this phenotype two replicate experiments were started from two different Oregon-K foundation stocks. In each experiment both an upward and a downward selected line were carried on. In the eighth and ninth generations of the upward selected line in experiment 2, and in the 29th generation of the up-selected line in experiment 1, flies occurred in which there was a slight enlargement of the halteres (i.e. a slight bithorax phenotype) in the normal environment. In both these cases the very slight degree of assimilation achieved was shown to be due to the presence of a single gene, namely an allele of *Bxl* with a dominant haltere-enlarging, and a recessive lethal, effect. In the 29th generation of upward selection in experiment 1, there also occurred some individuals in which, in the normal environment, the metathorax was more completely converted into a mesothorax. By 'second phase' selection among these, stocks were eventually built up in which the bithorax phenotype was very completely assimilated, being strongly expressed in over 80% of individuals in the normal environment.

The analysis of the genotype of this assimilated bithorax stock showed that it contained genes of two rather different types. On the one hand it was shown, by chromosome substitution experiments, that both the second and the third chromosomes contain factors tending to produce the bithorax phenotype in the individuals homozygous for them. Secondly, it was shown that there was present a sex-linked gene which exerted a recessive maternal effect, such that females homozygous for it lay eggs which tend to develop into bithorax phenotypes. If this gene is removed from the general background of the assimilated bithorax stock and crossed into a normal wild type, homozygous females still give rise to a few bithorax offspring, but the frequency with which the phenotype occurs is very much reduced. A somewhat unexpected finding was that, although females homozygous for the maternally acting gene lay eggs which tend, in the normal environment, to develop into bithoraxes, these eggs are no more sensitive than others containing the same general genotype to the effects of the abnormal environment (ether treatment shortly after laying), by which the abnormal phenotype was originally produced.

Summarizing these results we may say that in all cases in which complete or nearly complete assimilation has been achieved, the process has involved changes at many loci throughout the whole genotype. The only instances in which the genetic change was restricted, as far as is known, to a single locus are the two occurrences of a *Bxl*-like mutant in the bithorax experiments. In all the other experiments many loci must have been involved, but there is a considerable range between cases in which all the involved loci seem to be of relatively similar importance, to others in which one or a few loci are particularly strongly effective while the others can be considered as mere modifiers or genetic background. Thus, in the dumpy assimilation, a *dp* allele is so important as to be essential, though its action is modified by other parts of the genotype. In the bithorax assimilation a relatively important maternally acting sex-linked genetic factor can be detected, although attempts to determine whether the effect is produced by a single locus, or two or more loci, remained inconclusive; the condition, however, has by no means the same relative importance as the dumpy allele in the dumpy-assimilated stock. In the stocks in which various venation phenotypes were assimilated, it was, in general, impossible to identify any one predominant locus. Bateman made an estimate, for several of her stocks, of the relative importance of a hypothetical major gene as compared with that of the rest of the genotype in producing the assimilated character. For three of the assimilated stocks the estimates were that a major gene might be 1.57, 1.99, or 4.38 times as important as the remainder of the genotype. For a fourth stock the estimate was as high as 6.71, but this stock (known as *WE pcvl O*) had been derived in rather a special manner, and will be mentioned again later.

Origin of the genetic variation utilized in genetic assimilation

As we have seen, genetic assimilation fails completely, at least over periods of fairly small numbers of generations, in inbred strains lacking genetic variability. It can be successfully carried out when the foundation stock is a normal random-bred population in which considerable genetic variability is present. It is clear that this already-existing variability must provide a great deal of the genetic material which is finally incorporated into the fully assimilated stock. The question arises whether it provides all of it. It is difficult to see how to give, experimentally, a complete positive answer to this question, and we are driven to ask the somewhat less penetrating one, namely, is there any definite evidence that the assimilation process has utilized genetic variation which was not present in the foundation stock?

If the foundation stock contains all the genetic variation necessary for complete genetic assimilation, then we should expect to find, under a system of random mating, that the assimilated genotype would occur in the foundation stock under the normal environment although probably at an excessively low frequency. In Waddington's original experiments [1] with absence of posterior crossvein, no thorough search of the foundation stock for this phenotype was made, although it was not noticed in the inspection of a few hundred individuals. In Bateman's experiments the phenotype was found in 0.7% of one of the foundation stocks and 0.25% in another. Of the other phenotypes she investigated, the presence of an extra crossvein in the submarginal cell was found in 0.1% of the foundation, but the absence of the anterior crossvein and the excess crossvein in the first posterior cell were not found at all; neither was the dumpy phenotype. The bithorax phenotype was also not seen in the foundation stock. The failure to find these phenotypes in the foundation stock may, of course, only indicate that they depend on gene-complexes, the elements of which are all very rare so that their combination before selection starts is of excessive rarity.

From the population which showed 0.7% of posterior crossvein-lessness before selection started, Bateman selected and bred together the few crossveinless individuals. These responded gradually to selection, in a manner which indicated that many genes were involved, and it was not till generation 10 that the strain was producing 90% of crossveinless flies. This strain, which was produced by straightforward selection from the foundation stock, without the application of any abnormal environmental circumstances, was given the name *WE pcvl O*. It is interesting to note that when Bateman attempted to estimate the relative importance of a major gene as compared with the genetic background, it was in this strain that the major gene was of most importance, as has been pointed out above. It was presumably only because the foundation flies of this strain contained a comparatively powerfully acting gene that they exhibited the crossveinless phenotype in the background of the unselected foundation stock. The development of the strain must have involved the increase in frequency of the major-acting gene and also a concentration of other genes acting in a similar direction. In the other assimilated stocks which Bateman established from the same initial foundations, a greater part must have been played by the concentration of numerous genes acting in this direction and a lesser by increase in frequency of any single powerfully acting gene.

These results show clearly that, in these cases at least, the foundation stocks contained sufficient genetic variability to produce the strains in which the phenotype was assimilated. In other experiments in the series, however, there is evidence that new genetic variability, contributory to the constitution of the assimilated strain, has arisen during the course of the experiment. It seems that this must certainly have been the case for the two occurrences (the second of which may possibly have been due to a contamination from the first) of the *Bxl*-like allele in the bithorax experiments. Bateman also considered it likely that the dp^{TP} allele, on which the assimilation of the dumpy phenotype was based, arose by mutation during the course of the experiments. The mutation may, in fact, in this case, have been a result of the heat treatment since the first occurrence of a gene of this type was recorded by Plough (cf. [24]) in experiments on the heat induction of mutations. Finally, it appears rather likely that the maternally acting gene, which plays an important part in the final bithorax-assimilated stock, may also have arisen during the course of the selection, although the possibility cannot be excluded that it was present in very low frequency in the initial foundation stock.

GENETIC ASSIMILATION AND
DEVELOPMENTAL CANALIZATION
The canalization of development
Genetic assimilation obviously involves the somewhat paradoxical character of phenotypes to be, on the one hand, to some extent susceptible and, on the other, to some extent resistant, to alteration by environmental agencies. It is only because development can be modified by the environment that an acquired character can appear in the first place, when a population is transferred from environment E to environment E′; but when this character becomes assimilated, that implies that the development of the phenotype is now resistant to the effects of changing the environment back from E′ to E. The property of a developmental process, of being to some extent modifiable, but to some extent resistant to modification, has been referred to as its 'canalization' [25]. This notion can be applied whether the agents which tend to modify the course of development arise from genetic changes or from changes in the environment.

The fact that phenotypes are somewhat resistant to modification by changes in the genotype is a commonplace of genetics. Its simplest exemplification is perhaps in the phenomenon of dominance. In this, the alteration of one of a pair of dominant alleles to a recessive (from *AA* to *Aa*) does not succeed in bringing about any phenotypic alteration. The way in which geneticists have thought about this phenomenon was, for some years, directed on to rather inadequate lines by the assumption that dominance or recessiveness is a property of particular alleles. The origin of a more adequate view can be found in the work of Muller, Fisher, Stern, Goldschmidt, and others in the years around 1930. Fisher drew attention to the fact that the degree of dominance is subject to control by the rest of the genotype; and Muller discussed the phenomenon of dosage compensation, by which males containing only one dose of sex-linked factors appear very similar to females which contain two doses of the same factors, and emphasized that it is necessary, in such contexts, to consider the whole genotype and not individual alleles. The importance of the total genotype was also exhibited by the phenomenon of epistasis, in which inter-locus interactions lead to the concealing of certain genetic alterations. Again, genes of low penetrance are alleles whose action frequently fails to overtop some threshold set by the remainder of the genotype.

In all these cases, then, we are dealing with an essentially similar phenomenon, namely, a course of development which exhibits some resistance to being modified by genetic changes. It is, of course, also a well-known fact, but one with which in the past embryologists have had more to do than geneticists, that development also tends to resist being modified by environmental agencies. Embryos tend to regulate, that is to say, to produce their normal end-result in spite of external accidents which may occur to them as their development proceeds. Again, the existence of some resistance to modification is shown by the fact that different strains of the same species differ in the extent to which they are modified by a given external stress.

The notion of canalization is, therefore, intended to be a very general summing-up of a large number of well-known facts in genetics and embryology, all of which are summarized in the statement that the development of any particular phenotypic character is to some extent modifiable, and to some extent resistant to modification, by changes either in the genotype or in the environment. The concept was first formulated on the basis of a detailed study of the developmental effects of some forty genes which cooperate in determining the formation of the *Drosophila* wing; and at the same time it was pointed out that the fairly high degree of resistance to modification which is characteristic of the development of 'wild type' genotypes must have been produced by natural selection and is much reduced in many mutant forms[25–27].

Genetic assimilation as a consequence of canalization

Since the concept covers both types of possible modifying agent— genetic and environmental—it lends itself to a particularly easy exposition of the process of genetic assimilation, in which both these factors are involved. When a population is taken out of environment E and placed into environment E', the phenotypes developed by the different individuals will depend on the canalization of their developmental systems with respect to the environmental differences involved. If selection occurs for the production of certain acquired characters, this can be regarded as bringing about a change in the canalization determined by the genotypes that finally result from the selection. The canalization of development in the selected individuals will be such that the characteristic features of environment E' easily produce the acquired character in question. However, canalization, as we have seen, has two aspects: it involves not only the possibility of modification but also a resistance to modification. If the selected population is now placed back again into environment E, the resistance to modification of its phenotype may be such that this new change of environment does not modify the character back again to the form it had initially.

This way of expressing the situation is a highly general one, which can be applied either to characters whose development exhibits a threshold (such as the wing venation or bithorax phenotypes mentioned above) or those without thresholds (such as the anal papillae). Moreover, it does not attempt to specify by what mechanisms the resistance to genetic or environmental change is brought about. The genes which were present in the initial population, and which become accumulated by selection because they condition the appearance of the acquired character in the new environment, may have been concealed in the old environment either because, under those circumstances, they were of too small effect to be noticeable, or because they were hidden by dominance or epistasy.

Stern[28, 29] has offered what he considers to be an alternative explanation of the process of genetic assimilation, but it is difficult to see that his account differs in any way from that given above, except that it is expressed in slightly less general terms. Stern considers that the process is due to the selection of genetic differences which were 'sub-threshold' in the original environment E, but which are suprathreshold in the new environment E'. This is, in effect, merely another way of saying that, in the original population, the canalization of development is such that these genetic differences in the environment E remain concealed owing to phenomena such as dominance, epistasy, etc., while in the environment E' the environmental stress, added on to the effect of the genetic differences, allows them to come to expression. It is relatively easy to envisage such a situation when one is thinking in terms of canalization, that it to say, in terms of a course of development which has a certain resistance to modification by either genetic or environmental agencies. It is perhaps rather more awkward to do so in Stern's terminology, which leaves out of account the general process of differentiation, and concentrates its attention on the genetic factors, or the environmental factors, as separate agencies.

For instance, in discussing the application of genetic assimilation to the evolution of callosities on parts of the skin which normally are subject to pressure, Stern[29] expounds his 'new hypothesis' as follows: 'The hypothesis, in over-simplified form, begins with the fact that many genes have only a slight effect if they are present in single dose but a strong effect in double dose. If, for instance, the gene pair AA did not lead to the spontaneous production of specific callosities while the combination AA' implied a slight tendency towards this trait, then $A'A'$ might well cause its invariable appearance. In a population in which the gene A is highly abundant, and A' correspondingly rare, most individuals would be AA, a few AA' and practically none $A'A'$. Be it assumed that the AA' individuals will under pressure form callosities easier than AA individuals. They will then have been favoured by selection and their number will increase. The higher frequency will lead to crosses between AA' males and females, and as a necessary consequence of Mendelian segregation one-quarter of their offspring will be $A'A'$. This latter genotype is the very one which causes the spontaneous appearance of the callosities'.

This is, of course, exactly the same explanation (in oversimplified form) as has been offered above; but putting it like this leaves several loose ends. In the first place, it suggests that the concealment of the allele A' in the original population is due only to dominance, whereas it may also be due to inter-locus interactions or epistasy. In the second place, Stern has suddenly to bring into the middle of his argument what appears to be an extra *ad hoc* hypothesis when he writes: 'Be it assumed that the AA' will under pressure form callosities easier than AA individuals'. It is only because, in fact, everyone has at the back of his mind a vague idea, which it is the purpose of the notion of canalization to make explicit, that he accepts it as reasonable to suppose that one and the same gene has the two properties of having a slight tendency to produce callosities in a normal environment and increasing the tendency to react to pressure by the formation of callosities. Finally, the canalization terminology can deal with situations such as that of the anal papillae, in which no thresholds appear to be involved, and where some of the genes selected may very well operate simply to increase the extent to which the size of the papilla reacts to the presence of increased salt, without having any tendency themselves to cause a larger papilla when no salt is present. The existence of such genes, whose action is confined to the control of canalization (or reactivity, if one wishes to express it so), is not allowed for in Stern's scheme. They are, however, certainly a theoretical possibility and, as we shall see, there is considerable evidence for their existence.

Some concepts allied to canalization

1. *Autoregulation.* In the late thirties and early forties, Schmalhausen and a group of colleagues in Russia were following a line of thought almost exactly parallel to that which, at about the same time, led to the formulation of the idea of canalization and genetic assimilation. These ideas were brought together in a book published in Russian in 1947 and in American translation in 1949. Schmalhausen[30] discusses a notion very similar to canalization under the name 'auto-regulation', and he points out that the properties of the auto-regulating mechanisms of natural organisms have been brought into being by natural selection. He cites many examples of the apparent 'inheritance of acquired characters during evolution' and argues that they should find their explanation in selection for the capacity of organisms to react adaptively with their environment. He did not, however, formulate in any precise form the process which has been referred to here as genetic assimilation. Moreover, such experimental work of his school as is available in English seems to be related to the ideas nowadays referred to as the 'Baldwin effect' (see below), rather than to genetic assimilation. In any case, no experiments which were successful in bringing about a result which can be interpreted as genetic assimilation seem to have been reported by him or his students.

2. *Adaptive modifications and morphoses.* There are a group of notions connected with the nature of the response of an organism to an environmental stress which require discussion. Schmalhausen attempted to make a distinction between 'adaptive modifications'

and 'morphoses'. The former were, as their name implies, reactions to stress which were of adaptive value. Schmalhausen very convincingly pointed out that the capacity of an organism to react in such a way is almost certainly to be attributed to the past actions of natural selection. If, during the course of its evolutionary history, a given type of organism has frequently had to deal with a particular environmental difficulty, it is likely to have acquired, through natural selection, the ability to be modified by this stress in a way which is of use in coping with it. Schmalhausen used the term morphoses to refer to developmental modifications produced by abnormal environments which the organism has not encountered in its evolutionary past (e.g. high doses of ionizing radiation, etc.).

He seemed to imply that there was some real difference in the physiological nature on the developmental reactions involved in the two cases. It is, however, difficult to see why this should be so. If we consider the experiments on wing venation phenotypes described above, the reaction of a *Drosophila* population to a hot shock during the pupal stages must, in the first generation or so, be considered to be the production of a morphosis. However, under the influence of artificial selection, this behaves in exactly the same way as does the production of an undoubted adaptive modification, in the salt-treated larvae, when submitted to natural selection. There is no reason to suppose that, for instance, the adaptive modifications are under proper genetic control while the morphoses, in some way, escape from it; or indeed that the two categories differ in any matter of principle. It would seem likely that there is a continuous range, between developmental modifications in response to environmental stresses which have frequently been met in the past, and for which natural selection has built up a useful reaction, through those in response to circumstances for which natural selection has not been able to find an answer, to modifications produced by rare or unnatural stresses in connection with which previous selection has not been operative.

3. *Phenocopies.* Another term frequently employed in connection with the response of an organism to environmental stress is 'phenocopy'. This term was introduced by Goldschmidt[31], who has the merit of being one of the first to make a serious study of the reaction of organisms to external stresses, particularly those of a rather extreme character. Goldschmidt was primarily interested in throwing light, by such studies, on the mode of action of genetic factors. He felt that this could best be done by the study of situations in which some external factor, applied to a developing wild type individual, caused

the appearance of a phenotype which mimicked that produced by some recognized mutant allele. The word phenocopy was indeed coined to imply this copying of a mutant phenotype.

It has, however, gradually become clear that a knowledge of the nature of the environmental stresses which can produce a copy of a given mutant phenotype does not yield much information about the developmental action of the mutant alleles which produce similar phenotypes. The situation is, to employ the canalization terminology, that a particular phenotype is produced by a developmental process whose course is steered by the combined action of the whole genotype and the impinging environment, and that very often there are a considerable variety of genetic changes ('mimic genes') and of environmental stresses, any one of which will steer the development towards the production of a particular phenotypic abnormality. Moreover, when an abnormal environment succeeds in diverting development to some unusual end-result, it is not a matter of major importance whether a mutant allele happens already to be known which produces the same effect. For instance, in the venation experiments mentioned above the flies with broken posterior crossveins are phenocopies in the strict sense of the term, since a crossveinless gene —in fact several quite different crossveinless genes—are already known. The flies with extra crossveins in some of the anterior wing cells are, however, not phenocopies, since no gene producing this abnormality has apparently been described.

An important result of Goldschmidt's work was to show that different strains (in the early work, all wild type strains, but of different origins) differ in the frequency and type of developmental abnormality produced by the same applied environmental stress. This provided one of the bases for the notion that reaction to stress is a genetic property, and can be altered by selection. More recently studies have been made on the modifications caused by a given environmental stress when it is applied to strains which contain, either in homozygous or heterozygous condition, mutant alleles known to produce phenotypic effects similar to those which can be elicited by the environmental stress from wild type strains. For instance, Sang and McDonald[32] showed that sodium metaborate, which causes a reduction in eye size when fed to the larvae of wild type *Drosophila* strains at above a certain threshold concentration, gives the same type of effect with a lower theshold when fed to larvae heterozygous for the eyeless gene. These and similar results led Goldschmidt to pose the problem whether the phenocopying effect of a certain environmental stress could, in all cases, be regarded as the unmasking of a sub-threshold gene which was already present in the stock employed.

He devoted a good deal of work to attempting to find out the answer to this problem[33, 34], and showed that in many, though not in all, cases he could isolate some genetic factor tending to produce the phenotype in question. It may, however, be doubted whether the problem is a real one. A phenocopy (or, in general, any abnormal phenotype elicited by an environmental stress) must result from the combined action of the environment and the geno-type of the organism. That is to say, genes must be involved in the production of phenocopies, and *ex hypothesi* they must be sub-threshold genes, at least as regards the abnormal phenotype. The only question with which Goldschmidt was really concerned was whether, in the cases he investigated, there was any one gene which was of sufficiently great importance to be identifiable as *the* sub-threshold gene. The situation is strictly comparable to that in the assimilation experiments described above, in which, as we saw, it was sometimes possible to identify a single relatively important locus, and sometimes not. As Landauer[35, 36] has put the matter, in discussing rare developmental abnormalities ('pheno-deviants'): 'Our evidence leads us to conclude that sporadic defects, as well as experimental phenocopies, are the results of events through which ordinarily hidden weaknesses of developmental equilibria become manifest, and that these weaknesses have a definite, if complex, genetic basis. If the phenocopy concept in its narrow meaning of a purely environmental interference with developmental processes must be abandoned, it is clear that the existence of crypto-genes and their spontaneous or experimental aberration confront us with many new problems'.

Landauer's phrase 'developmental equilibria' is, of course, a short-hand form of referring to the notion of canalization, but the word 'equilibrium' is not quite satisfactory in this context for two reasons (cf.[37]). Firstly, the idea of equilibrium tends to be an absolute one, whereas we need to discuss just how well equilibrated, or stable, developmental processes are (cf. Landauer's reference to 'weak-nesses' of equilibria). And secondly, developmental processes are not in equilibrium in time, but essentially change as time passes. Both these points are easily dealt with in the canalization terminology.

4. *Homeostasis.* Another word which has been introduced into these discussions, and requires some notice, is 'homeostasis'. The classic use of this term in biology is by Cannon[38]. He used it to refer to the fact that many physiological systems show a capacity to return,

after disturbance, to some standard condition in which they maintain themselves. For instance, if the pH or CO_2 tension in the blood is altered in some way, processes come into operation which tend to annul the alteration and restore the previous condition. Such a situation can be referred to as one of 'physiological homeostasis'. More recently the word has been used in at least three other different contexts, in addition to that of physiology.

Lerner[39] has used it to refer to the tendency of a population to maintain, and if necessary restore, a particular distribution of gene-frequencies within it. If, for instance, the gene-frequencies in a population are altered by selection for a few generations, and the population is thereafter left to itself with no further selection, the gene-frequencies are frequently found to return to their natural values, the operative machinery being largely that of natural selection. These and other similar phenomena he referred to as 'genetic heomeostasis'.

This usage is rather far from our present context, and need cause little confusion in it. The situation is rather different concerning the other two usages, in which the word homeostasis is used in conjunction either with developmental or with general evolutionary processes. The word presumably always implies that something or other is being held constant, but when it is used in connection with evolution and development it is often difficult to decide what exactly this constant element is supposed to be. The natural interpretation of such a phrase as 'developmental homeostasis' would be that development is being held constant, or to put it more precisely, that development is tending to reach a constant end-result. However, this is not what authors who employ the phrase always have in mind. For instance, Lewontin[40] uses the phrase 'developmental homeostasis' to mean developmental mechanisms (including those which lead to an unusual end-result) which prevent the evolutionary fitness of an organism from being reduced in the particular environment in which it finds itself.

The tendency to keep fitness constant, which Lewontin[40] and Dobzhansky[41] have referred to simply as 'homeostasis' (without qualification) could perhaps be usefully referred to as 'evolutionary homeostasis', although the phenomenon is really a maximization of fitness rather than holding it constant, since if in a given environment fitness increased, there would be no tendency to reduce it. In the context of evolution this 'homeostasis' of fitness is of overriding importance, and natural selection will tend to produce some mechanism for bringing it about. Such mechanisms may involve holding the

end-product of development constant ('developmental homeostasis')
or they may involve the almost precisely opposite gambit of allowing
development to be altered in an adaptive way[42]. The use of the
word 'homeostasis', particularly if not qualified in a way which
definitely states what aspect of the animal's existence is being held
constant, therefore almost inevitably leads to confusion. The
canalization terminology on the other hand is designed precisely to
deal with the complementary flexibility and inflexibility of developmental processes which natural selection exploits to ensure that
organisms can keep their fitness maximal in different environments.
Further discussion of these terminological points will be found in
Waddington[9].

Dobzhansky and his colleagues have recently made many very
important studies on the genetic systems in wild populations and their
response to different environments[41, 43]. Some aspects of the
phenomena which they have revealed, and have discussed in terms
of homeostasis, are certainly related to what we have called developmental canalization, and might perhaps be less ambiguously
expressed in that phraseology. This work does not, however, have a
sufficiently direct bearing on genetic assimilation for it to be in place
to attempt such a discussion here.

Another attack on similar problems, in which the physiological
and genetic mechanisms have been, perhaps, more definitely
integrated, has recently been started by Forbes Robertson[44, 45].
He has studied the effect of selecting a strain for a certain character
(usually body size) in one environment on its developmental
reactions and fitness when taken into some other environment. The
environments used were rather precisely known nutritional conditions, and the growth performances of the organisms were analyzed
in terms of cell numbers, cell size, and developmental time. The
results clearly exhibit both the flexibility and inflexibility of development which constitute canalization, and the effects of previous
selection on the ways which the organisms adopt to maximize their
fitness under different circumstances. Another study in which the
technique of selection has been used to investigate the genetic constitution underlying responsiveness to environmental stresses is that
of Falconer[46]. In both these pieces of work, however, the results,
although very pertinent to the concept of canalization, are not
closely enough related to genetic assimilation to be discussed further
here.

The genetic basis of canalization
There are a few experiments which have explicitly set out to investi-

gate the genetic basis of canalization and which require at least short discussion.

The idea of canalization involves no more than that the course of development exhibits, in some way, a balance between flexibility and inflexibility. The particular form that this balance takes may vary widely in different cases [47]. The character of the canalization in a particular case can be expressed by plotting the extent of developmental response against the magnitude of the disturbing stress, either genetic or environmental. We may find cases in which the dose-response curve is linear; others in which variations in dose near the normal range produce considerable effects, but it becomes more and more difficult to alter development as it diverges from the norm (this is probably the case for characters such as body size); or other cases in which considerable variations around the normal dose remain almost ineffective, while with still greater doses the development becomes more easily deformable (this is so when the normal phenotype is protected by thresholds or quasi-thresholds). Most of the experimental analyses of canalization have dealt with cases of the last type.

1. *Variation around the wild type.* The first such study [48] was designed to discover whether the normal phenotype of the *Drosophila* wing venation pattern is protected by thresholds, i.e. is a case of canalization of the last of the three types mentioned. Four stocks were prepared, in two of which there were absences (either major in one stock, or minor in the other) of the posterior crossvein, while in the other two there were excess pieces of vein (either small or large) attached to the posterior crossvein. A series of crosses, backcrosses, and so on were made between these stocks; and the conclusion clearly emerged that the wild type phenotype can conceal within it a much greater range of dosages of vein-producing genes than can any other phenotype. That implies that the canalization of the normal vein pattern is such that it is highly resistant to the disturbing effects of changes in the dosage of genes tending either to make more or to make less vein. In individuals in which the buffering capacity of the normal developmental course is exceeded, the phenotype does become altered, and in these the phenotype which develops reflects not too inaccurately the actual dosage of genes contained in them.

The situation was actually somewhat more complex than this, since the character 'quantity of posterior crossvein' turned out not to be 'causally homogeneous'; that is to say, the same total quantity of vein could be produced in different ways, for instance, by the presence of a single complete crossvein or by the presence of a broken crossvein plus the addition of an extra fragment attached to its side. A similar complexity arises in some of Forbes Robertson's work, where the same body size can be achieved either by an increase in cell number or an increase in cell size. In such cases the global character which is under investigation is made up of more than one constitutive developmental course, and each of these requires consideration on its own. It remains true, however, in the case of the crossvein experiments under discussion, that each of these developmental courses exhibits canalization of the kind described.

2. *Disruption of canalization.* When a wild type phenotype is strongly canalized very little of the genetic variation present in a normal population will come to expression and thus be available for selection. A greater amount of the variation can be revealed if some way is found to push the processes of development away from the canalized phenotype, so that they follow a path which is more susceptible to the influence of minor genetic variation. Development can be steered into a less well canalized pathway not only by environmental stresses, as in the experiments on genetic assimilation, but also by the influence of some major gene affecting the character under investigation. One can therefore use such genes to destabilize the development of a phenotypic character which then becomes more responsive to selection.

Dun and Fraser [49], Fraser *et al.* [50], and Fraser and Kindred [51] have used this method to carry out selection on the number of vibrissae in the house mouse. The number of secondary vibrissae in the mouse is normally 19, and in most stocks there is very little variation about this number. The gene Tabby reduces the number to about half, and, the Tabby phenotype being less well canalized than the wild type, there is considerable variation. Dun and Fraser kept a line in which Tabby (which is a semidominant) was segregating in each generation. They selected for either increased or lowered number of vibrissae in the Tabby heterozygotes, and found that there was a parallel increase or decrease in numbers in the wild type sibs; the changes produced in the wild type were, however, very much smaller than those produced in the Tabbies. Thus, selection which changed the vibrissae number in Tabby heterozygotes from about 11 to about 16.5 produced an increase of only about half a vibrissa in the wild type sibs. This result shows clearly that the wild type phenotype is much more resistant to modification, that is, is much more strongly canalized, than is the Tabby phenotype.

Rendel [52, 53] and Rendel and Sheldon [54] have conducted

very similar experiments in which they used an allele of scute to force the development of the scutellar bristles in *Drosophila* out of their normal canalized path. In the wild type there are 4 scutellar bristles, and usually very little phenotypic variation away from this number. The scute gene in the original population reduced the mean number to 2, and there was a good deal of variation around this. Selection for scutellar number in lines segregating for scute raised the number gradually to 4, and at the same time the number of bristles in the wild type sibs increased. Again, the change in the wild type was much smaller than that in the destabilized scute stock. Rendel suggested a method of estimating the relative genetic change required to bring about different changes in bristle numbers; and he calculated that to move from 3 bristles through the canalized number of 4 to 5 bristles takes about eight times as much genetic change as it does to move from 1 bristle to 3. This gives an indication of the degree to which the canalized normal phenotype can absorb or conceal genetic variation.

3. *Improvement of canalization*. Rather similar work has been published by Maynard Smith and Sondhi[55] concerning the number of ocelli which in *Drosophila* is normally canalized at three. Maynard Smith[56] has also discussed the degree of precision which must be achieved by the underlying physical processes of the canalized system in order to ensure the constancy in number of organs such as ocelli and bristles.

Another method of demonstrating the reality of canalization is to produce changes in the degree of canalization by means of selection. Wild type phenotypes are already so well canalized that it is not easy to alter their degree of canalization. This can, however, fairly easily be done in the relatively destabilized phenotypes produced by mutant genes. For instance, in *Drosophila melanogaster*, Bar causes a considerable reduction in the size of the eye, and it is well known that the Bar phenotype is sensitive to temperature, the eyes being larger at low temperatures and smaller at high temperatures. Waddington[57] practiced selection for increased canalization by the following method: a number of pair matings were set up, the progeny being divided into two lots reared at 18° and at 25°, and selection made by taking those families in which the difference in eye size at the two temperatures was least. After some six generations, families were obtained in which the difference in facet number at the two temperatures had been reduced to between 6 and 15 facets, whereas in the unselected foundation stock, tested at the same time, the difference was about 100 facets. Thus a very considerably increased degree of canalization against the disturbing effects of temperature had been achieved.

4. *The genetic structure of canalized phenotypes.* Many aspects of the wild type phenotypes are not so well canalized as the number of scutellar bristles or the number of ocelli. The application of the idea of canalization to phenotypes which exhibit continuous variation in normal populations was discussed by Waddington at a conference which took place in 1950[47]. A few years later, Lerner[39] suggested that the homeostasis of gene-frequencies in a population often depends on heterozygotes exhibiting better developmental canalization than homozygotes and therefore being biologically fitter. Several authors have studied Lerner's suggestion that heterozygotes exhibit better developmental canalization. Most of these investigations have used, as an indication of canalization, not the response of the phenotype to some known exterior stress, but the variation, within the body of a single individual, of some phenotypic character which is repeated in a number of different parts of the body which might be expected to be identical.

For instance, Mather[58], Jinks and Mather[59], Tebbs and Thoday[60], Thoday[61, 62], and Reeve[63] have studied the degreee of asymmetry in numbers of sternopleural bristles on the two sides of a fly, while Reeve and Robertson[64] have considered the correlation in numbers of abdominal bristles on the various sternites of *Drosophila*. The results make it clear that there is some genetic influence on the degree of asymmetry or repeatability of such characters, but the heritabilities found were in most cases very low. Heterozygotes often exhibited a lower degree of asymmetry, or a higher degree of repeatability, than the corresponding homozygotes from which they were derived, but this rule was by no means always the case.

Doubts have been expressed, however, (e.g. [9, 64]) whether the character investigated in these studies, which is the extent of the phenotypic variation produced by intangible alterations of conditions during development, is really an indication of the degree to which the developmental systems are canalized against more definite environmental changes or alterations in genotype, and whether they can be taken to throw much light on the genetic make-up of canalization systems in general. It is worthy of note that in Waddington's [57] experiments on the Bar phenotype, the lines showing a high degree of canalization were considerably more inbred than the foundation population. It is therefore clear that a high degree of heterozygosity is certainly not necessary for strong canalization, which can be achieved in relatively homozygous genotypes if selection is made for it. In these same experiments it was also found that a high degree of canalization against the effects of external temperature was not accompanied by a reduction in the asymmetry of the eyes on the two sides. This strongly suggests that the variation between parts which are repeated within the same individual is not a good indication of canalization in general. Waddington[9] suggested that it should be referred to as 'developmental noise'. Clayton *et al.*[65] have spoken of it as 'developmental error', and Reeve [63] refers to it as 'chance variance'.

In the development of any complex phenotype there will be a large number of developmental pathways leading to the various final adult organs or characters. Mather[66] at one time attempted to discuss the organization of such a developmental system in terms of a proposed distinction between 'oligogenes' and 'polygenes'. The former are the well-recognized and distinct alleles with which genetics has been mainly concerned, while the latter were supposed to be genes of a special kind, possibly constituted by heterochromatin, which produce only small effects on the phenotype and which were thought of as responsible for the production of continuous variation. Mather suggested that the canalization of the various pathways of development is dependent on polygenes, while oligogenes act mainly by switching parts of the developing system into one or other of the different paths. However, Waddington[67] argued that the distinction between polygenes and oligogenes is not a real one, since the same allele can appear as a polygene in one genotype but as a well-recognizable oligogene in some other genotype. In this particular context he claimed that there was no reason to suppose that the switching between one path and another could not be done by systems of genes each of small effect (i.e. Mather's 'polygenes'), as well as by single identifiable alleles. This has since been shown to be the case, as in the crossveinless experiments described above.

Fraser[68] has shown how the conditions for the evolution and maintenance of genotypes determining canalized paths of development can be investigated by Monte Carlo methods on a digital computer. This method seems likely to become a powerful tool for studying the complex theoretical structure of such situations, but so far no very striking results have appeared.

5. *The evolution of canalization.* Berg[69] has given an interesting discussion of the factors which will lead to the evolution of high degrees of developmental canalization. He argues that this will occur when the aspects of the environment which exert natural selective

pressure on an organ have, in themselves, no direct effect on the developmental processes by which that organ is produced. It will then be necessary for the developmental system to produce an organ which is precisely tailored to meet the requirements of natural selection, and to do so without being able to use the selective factors to guide the process of development while it is actually going on. As an example, Berg quotes the very low variance (i.e. high degree of canalization) of the dimensions of insect-pollinated flowers which have to deposit their pollen on to particular portions of an insect's anatomy. Here the selective force, i.e. the correct placing of the pollen, certainly plays no part in the developmental processes by which the flower comes into being. It is for this reason, Berg argues, that it is necessary for the development to be very highly canalized, so that flowers of exactly the right dimensions are reliably produced in spite of variations in temperature, moisture, and other climatic conditions during the growth of the plant.

While there are many cases of canalization to which Berg's explanation may be applied, there are probably others which do not so easily fall within its scope. The most difficult fact to understand about the evolution of canalization is its tendency to, as it were, overshoot the mark which would seem to be necessary. Even if one grants that all developmental processes have a certain quality of inflexibility, and that selection for a comparatively extreme type of developmental modification will, owing to this inflexibility, lead to some genetic assimilation of the character in the way described above, it remains rather peculiar that natural evolution so frequently results in a much firmer genetic assimilation than would appear to be demanded. For instance, selection for the ability to form thickened skin on the soles of the feet as a response to pressure might well be expected to lead to some degree of assimilation of the character, so that the foot soles were thick even in people who never walked. Why, however, should this assimilation go so far that the foot soles thicken even *in utero*, before there is any possibility of their being used? One may guess that some sort of 'factor of safety' consideration is coming into play, but it is not clear why it should operate in such an apparently exaggerated form.

One possible mode of approach to the topic would be to draw attention to the fact that development is usually easier to modify at early stages than late. It may be, as it were, easier for the genotype to produce an allele which causes thickening of the soles of the feet at an early embryonic stage *in utero*, rather than one that operates exactly at the time when the thickened skin becomes advantageous.

We still understand so little about the actual processes involved in development that it is difficult to know how much weight should be lent to such consideration. But we are perhaps too ready to believe that it is quite easy to produce a gene that will do anything required. Actually it may be that mutation plus natural selection should only be expected to produce a system which 'gets by', rather than one that does exactly what is required and no more. It is noteworthy that in the bithorax assimilation experiments, one of the important genes involved was a maternal-effect gene which operates on the egg out of which the bithorax individual will develop; that is to say, it acts very much earlier in development than the period at which the selective pressure was exerted. If the species has to use whatever genes it has at its disposal to meet the demands of natural selection, it seems quite likely that it will have to use genes which act earlier than necessary, since it cannot make use of those which act later.

It is presumably mechanisms of this sort which have brought about the great evolutionary divergence in the types of eggs characteristic of different phyla and families of the animal kingdom, since it is very unlikely that these can have been shaped by natural selective forces acting at such early periods of ontogeny. The genes which determine egg structure have probably been fixed by natural selection acting on the adults to which these eggs give rise, rather than on the eggs themselves.
6. *The 'tuning' of canalized pathways.* It seems reasonable to expect that the genetic control of canalized processes of development will affect not only the readiness with which a process is modifiable by an environmental stress, but also the precise character of the response, and thus of the end-state to which development eventually leads. In so far as this is so, natural selection will tend to build up genotypes which not only respond to the environment, but do so in such a way as to produce a phenotype which has positive selective value.

That such theoretical predictions may be realizable in practice is demonstrated by Bateman's experiments on *Drosophila* venation described above. When the environmental stress (heat shock) was applied to a normal wild type stock, a number of different phenotypic modifications were produced; and when any one of these (e.g. an extra crossvein in the submarginal cell) was chosen to be favoured in selection, genotypes were rapidly produced which gave rise to that modification rather than any of the others. To use somewhat picturesque language, one might say that the selection did not merely lower a threshold, but determined in what direction the developing system would proceed after it had got over the threshold.

The determination of the direction taken by a canalized pathway has been referred to as the 'tuning' of it[9]. The phenomenon is likely to be of particular importance for evolution when one is dealing with highly complex environmental stresses, such as those which might arise from some new manner of bodily movement, such as, to take a relatively simple example, the digging movements practised by burrowing animals. If the responses to such stresses are to be of adaptive value, it is essential that they shall take place in a whole series of organs in a co-ordinated way, and this in its turn requires that the modification of each individual organ shall be precisely determined so as to fit in with the over-all pattern. In a normal above-ground mammal which takes to burrowing, the responses of its claws, fingers, arm muscles, shoulder girdle, and so on to the stresses involved will have to be delicately related to one another if the whole system is to be efficient.

In any given species, at a particular time in evolutionary history, much of this correlation is probably brought about by basic developmental mechanisms. In vertebrates at the present day, for instance, there seem to be mechanisms which ensure some basic congruity between the skeleton and muscles of an organ such as the limb, as is shown by the fact that polydactyly genes, for instance, usually produce not only extra digital bones, but the extra muscles to go with them. This is, theoretically, by no means a necessary state of affairs. For instance, in bithorax phenotypes of *Drosophila*, the metathorax which is transformed into a mesothorax does not always contain the appropriate mesothoracic muscles[70]. These mechanisms, when they exist, must have been brought into being by natural selection acting in the long-past evolutionary history of the organisms, but it is not intended to discuss such 'macro-evolutionary' processes in this article. However, even in an organism provided with developmental mechanisms which produce some correlations between systems which must function in harmony, something more will be necessary if the evolving organs are to be maximally appropriate for the tasks they have to perform. For instance, a developmental mechanism which ensured some degree of correlation between a limb and its girdle could hardly be expected to produce exactly the right modification of the girdle and its musculature to suit a limb strengthened by intensive digging activities. The final 'tuning' of the adaptive modification will have to be carried out by a selection of genes which control the minor details of the correlations.

THE BALDWIN EFFECT

Around the beginning of this century, Baldwin in America and Lloyd Morgan in England made suggestions about a type of evolutionary process which might have results very similar to those of genetic assimilation. They spoke of the process as 'organic selection'. After the rediscovery of Mendelism and the application of Mendelian ideas to evolutionary theory, the idea of organic selection fell rather into the background, and was scarcely discussed at all until Huxley devoted some ten lines to it in his large work of 1942 [71]. It was not until shortly after the first experimental work on genetic assimilation was published that Simpson [72] brought it to the forefront as an alternative way of accounting for apparently Lamarckian inheritance of acquired characters.

Baldwin's and Lloyd Morgan's discussions were, of course, couched in pre-Mendelian language, and it is not entirely easy to see exactly what their meaning would be when translated into terms of our modern concepts. Most of the authors who have referred to the subject recently, however, seem to understand the theory of organic selection to be that organisms may be able, by nongenetic mechanisms, to adapt themselves to a strange environment, in which they can then persist until such time as random mutation throws up a new allele which will produce the required developmental modification. Natural selection will then increase the frequency of this new allele, so that the developmental modification, which was originally an acquired character, will become an inherited one.

The process, if understood in this sense, differs from the notion of genetic assimilation primarily because it considers the initial adaptation to the new environment to be a nongenetic phenomenon on which selection has no effect. Thus Mayr [73] speaks of this adaptation as being due to 'a nongenetic plasticity of the phenotype', and Huxley and Simpson use essentially similar phrases. This implies that natural selection is having no effect on the system until such time as a new allele appears which causes the production of the acquired developmental modification.

This theory seems to be an impossible one [74]. The acquirement of an adaptive modification in response to an environmental stress cannot, according to all our basic ideas of genetics, be due simply to a plasticity of the phenotype to which the genotype is quite irrelevant. The adaptive modification, like all other characters of the developed animal, must be an expression of the hereditary potentialities with which the zygote was endowed. Moreover, as we have seen in the genetic assimilation experiments described above, these theoretical considerations are borne out in practice. The ability to acquire a character is a genetic property of a strain, and can be modified by selection. If a population persists in some unusual environment by forming a suitable adaptive modification, natural selection is bound to operate on the genetic factors which control its ability to react in this way. The situation must, in fact, be one of the kind contemplated in the theory of genetic assimilation. The idea of organic selection, at least as it has been interpreted recently, cannot be accepted as a possible alternative to the genetic assimilation mechanism. It is merely an out-of-date speculation which should be allowed to lapse back into the oblivion from which Huxley and Simpson rescued it.

Only very little experimental work has been done in which the notion of organic selection was taken as a guide. However, one series of experiments was carried out by Naumenko[75], a pupil of Schmalhausen, which should perhaps be considered to fall in this category. Naumenko produced developmental modifications in *Drosophila* by administration of a heavy dose (about 4000 R) of X-rays. He states that he used this agent because it is known to produce developmental abnormalities and because it is a mutagenic agent which would speed up the production of random mutations. Naumenko continued the treatment for about a dozen generations on each of two lines, in one of which he selected for the development of rough eyes and in the other for that of divergent wings. In both lines there was, according to his figures, a slight increase in the frequency with which these effects were produced.

Naumenko, however, does not appear to have been looking for a gradual response to selection for ability to respond to the environmental stress in these particular ways. He was expecting to find that definite and identifiable mutations would occur which would produce rough eyes or divergent wings, and which would then be fixed by the selection. In point of fact, the result he obtained was that in both lines, after some time, mutations producing divergent wings appeared and could be fixed by selection. This happened in the stock which was being selected for rough eyes as well as in that being selected for divergent wings. Naumenko concluded that the mutations had arisen independently of the selection of the flies for the developmental modifications.

He seems to feel, nevertheless, that the experiments gave grounds for believing that a 'Baldwin effect' is a theoretical possibility. One might perhaps concede that, when one is dealing with such an extremely abnormal environmental stress as high-intensity irradiation, the process of genetic assimilation is likely to be very ineffective; one can scarcely imagine that normal populations carry much genetic variation affecting the way in which an organism's development is modified by such drastic and unnatural agencies. Selection for an appropriate type of response could, therefore, not progress very far, and the only chance for the genetic fixation of the character would be the occurrence of a new mutation. Such types of environmental stress, however, do not provide a very convincing argument for the processes which may be expected to proceed in natural evolution. So far as Naumenko's experiments validate the idea of organic selection at all, they do so only by suggesting that it might be an extreme limiting case of genetic assimilation, occurring when selection among the genes pre-existing in the population is least effective, and almost all the effect has to depend on the occurrence of new mutations. [For further consideration of the Baldwin effect see p. 279]

SUMMARY AND GENERAL CONCLUSIONS

The process of genetic assimilation is one by which a phenotypic character, which initially is produced only in response to some environmental influence, becomes, through a process of selection, taken over by the genotype, so that it is formed even in the absence of the environmental influence which had at first been necessary.

The occurrence of such processes has been demonstrated in *Drosophila*, both for characters whose development involves thresholds and for others which do not.

Genetic assimilation is brought about by the operation of orthodox genetic and embryological principles. It depends on two main types of fact: (a) that the capacity of an organism to be modified in response to an environmental stress is under genetic control, and can be altered by selection; and (b) that developmental processes exhibit a balance between tendencies to be modified by the environment and tendencies to resist modification.

This balance between flexibility and inflexibility can most easily be expressed in terms of the concept of 'the canalization of development'. This notion can be considered to follow from general principles; and special investigations of various theorems deduced from it have shown it to be fully justified, but have not shown that it depends on any single type of genetic system, such as, for example, heterozygosity.

An alternative mechanism for bringing about similar results is the 'Baldwin effect'. This supposes that the responsiveness of the organism to environmental stress is not under genetic control, and it therefore does not invoke the operation of selection in converting an 'acquired' character into an 'inherited' one, but suggests that this occurs as a consequence of a chance mutation. It is argued that the rejection of the genetic control of responsiveness and of the operation of selection is contrary to general genetic principles and cannot be accepted. The 'Baldwin effect' is therefore not a possible general mechanism of evolution; it is, at most, no more than the limiting case toward which genetic assimilation tends when the operation of selection of the genetically controlled capacity to respond is minimally effective.

The main conclusions to be drawn are, not that any new fundamental genetic principles have been disclosed, but rather that it has been shown that well-accepted principles lead to evolutionary consequences quite other than those which have usually been supposed to follow from them. It has been conventional to argue, from the fact that 'acquired characters' are not inherited, that the development of adaptive modifications during the lifetime of an organism is irrelevant to the evolution of similar genetically determined phenotypes. The theory of genetic assimilation, and the practical demonstration that the process can occur, shows that this argument is misguided, and provides a new way of accounting for all those evolutionary facts for which, in the past, some authors were tempted to advance a 'Lamarckian' explanation. The explanation in terms of genetic assimilation is not alternative to an account in terms of random gene mutations, but is supplementary to it; and what it adds to that theory is the more important, the more one is dealing with evolutionary changes in complex organs or organ systems which must be affected by numerous genes which have to be integrated with one another.   [1961]

GENETIC ASSIMILATION IN LIMNAEA

The most thorough and interesting study of genetic assimilation under natural conditions concerns the well-known pond snail *Limnaea stagnalis*. It was made by the great Swiss psychologist Piaget, and was begun in 1929 before he had taken up the study of psychological development in man for which he has become so well known.

Limnaea, like other gastropods, produces a shell which is laid down continuously throughout its life, and which therefore serves as a record of its whole developmental history. Species of *Limnaea* in general are famous for their variability of phenotype. *L. stagnalis* has normally an elongated shape, as shown in figure 1 (a). However, in some of the large Swiss and Swedish lakes it produces a variety of a shorter kind known as *lacustris* (figure 1 (b)), and in Lakes Neuchatel and Constance there is a still shorter variety (*bodamica*,

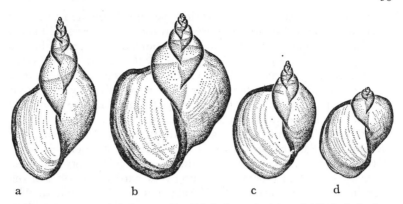

Figure 1. Limnaea. (a) *L. stagnalis,* (b) *L. lacustris,* (c) and (d) *L. bodamica.*

figure 1 (c) and (d)). In nature these shorter varieties appear only in fairly shallow water, where the lake bed is flat, stony, and much exposed to wind and wave action. Piaget was able to show experimentally that similar phenotypes can be produced as physiological adaptations to wave action. The young snail emerges from its egg with only one or two gyres on its spiral shell. During the growth period, until it attains the adult form with six or seven gyres, snails (at least of the more or less contracted races) tend to contract their columellar muscle to hold themselves firmly to the bottom whenever a wave or disturbance threatens to dislodge them, and as a consequence of this continued contraction the shell develops with a shortened form.

When contracted forms are collected in nature and brought into the laboratory they retain their contracted phenotypes through many generations, even when bred in still water in laboratory tanks. Piaget also transplanted a number of eggs belonging to the most contracted form (taken from the sixth generation, reared in the laboratory) to a still, natural pool in which there had previously been no *Limnaea stagnalis.* Again, in this situation they survived for a number of generations without losing their contracted phenotype. One must conclude, therefore, that the contracted phenotype has been definitely assimilated. The form, which had originated as a physiological reaction to the stress of wave action, has been converted, presumably through selection, into a genetically assimilated condition, which develops independently of a particular precipitating stress.

Contracted races studied by Piaget were characterized not only by the shape of their shell but by certain types of behaviour. In the first place, they 'choose' their environment in the sense suggested below. They are, of course, quite free to move about the lake, and there is nothing to prevent them migrating into greater depths where wave action is no longer crucial. However, there is no sign at all that they do so. In Lake Neuchatel the deeper water is inhabited by a different race, known as *Bollingeri*, which are quite small in size and very elongated in shape, and at intermediate locations there is a strictly bimodal distribution of shell types, with little sign of intermediates. Moreover, the contracted types show a different type of behaviour in response to disturbance when they are in laboratory cultures. Normally elongated types do not cling very hard to the sides of the aquarium jars, and are easily dislodged if the jar is jolted. The contracted types, however, cling much more tightly, and it is very difficult to dislodge them by merely jarring their container.

We have here a good example of the first of the evolutionary phases outlined in the upper part of figure 9 of paper 7, on page 53; namely, habitat selection and special forms of behaviour.

In discussing these *Limnaeas* in his recent book, Piaget at first accepts the reality of the process of genetic assimilation, which of course had not been proposed at the time he began the work. He goes on to argue that although this has almost certainly occurred it is not in itself sufficient to explain all the facts. First, he points out that, even in populations of contracted forms, a contracted shell shape is not fully genetically fixed in all individuals, and when such a population is brought into the still waters of laboratory tanks some of the next generation grow up with little or no contraction. He concludes that phenotypic adaptability alone is sufficient to produce an adequate contraction, and he wonders, therefore, what role selection can play. This argument seems to overlook the fact that there will be genetic variability in the capacity to perform the physiological adaptation which brings about the contraction. It is only to be expected that in a population, the adult phenotype in some individuals will have been produced with a stronger influence of the environment, and in others with a stronger influence of the inherited genotype. It is only with a strong artificial selection, or an exceptionally long period of natural selection, that one could expect to produce a population which was homogeneous with respect to the contributions of genotype and environment to the production of the adapted phenotype. This observation of Piaget's therefore seems to me to have little force.

Paiget's second argument is rather similar. It concerns the origin of the genetic differences between the varieties *lacustris* and *bodamica*. Of these two, *bodamica* can be said to be the genetically more contracted, in the sense that when populations of both varieties are brought into the laboratory and thus removed from wave action, *bodamica* develops in a more contracted form than *lacustris*. This is in fact, the main criteria for distinguishing the races, since in nature, under the action of waves, individuals of both races may develop indistinguishably contracted shells. The variety *lacustris* occurs in the three lakes, Geneva, Constance, and Neuchatel, while the most contracted race, *bodamica*, occurs only in the last two which are more highly disturbed by wave action. Piaget poses the question why, if *lacustris* with sufficient wave action can produce a race able to survive in the rough lakes Constance and Neuchatel, was there any selection pressure to force *bodamica* to produce a genotype causing even stronger contraction? He feels that these facts are evidence not only of selection of genes controlling adaptability, but also a more direct Lamarckian-type action of the continued environmental conditions, bringing about actual alterations of the genotype by some mechanism which he can only specify very vaguely as a 'progressive reorganization, or gradual change in the proportion of the genome'.

The situation to which Piaget draws attention is one that is continually met in experiments on the selection for phenotypic adaptation. If one selects for a given phenotypic modification (e.g. a break in a *Drosophila* cross-vein as a response to heat shock), and finds that the phenotypic modification eventually becomes genetically assimilated so that it can occur without the precipitating environmental stimulus, the reason must be that there are many genes tending to produce the modified phenotype under the influence of the suitable environmental stress. The selection gradually brings together many different genes tending in this direction, and the combination of these genes suffices to produce the phenotype, even with the environmental stresses removed. In Piaget's example, variety *lacustris* can produce sufficiently contracted forms under the conditions of Lakes Constance and Neuchatel. One has to suppose that there is some genetic variation between individuals of *lacustris* in their ability to do this. Selection will therefore tend to accumulate a number of *lacustris* genotypes with a tendency to develop in this manner. It is a collection of genotypes of this kind that forms the race *bodamica*. In a way, one might say that selection has done more than it need to, but this 'making doubly sure' is characteristic of the process of genetic assimilation.   [1973]

10

Unpublished correspondence between C. H. Waddington and
Professor Th. Dobzhansky, Columbia University, New York,
during July and August 1959.

Waddington to Dobzhansky, 25 July 1959

I have just received a lot of reprints from you, which I found very
interesting, particularly one on homeostasis and senescence. As a
matter of fact, I have just been trying to do an experiment on this. I
tried to (1) take a stock of Bar, and improve its canalization against
temperature differences, as in the experiments I have recently published in *Nature*, (2) see whether this buffered strain was less affected
in its life-span by a sublethal temperature treatment. The experiment
is proving pretty difficult to carry out, and so far we haven't made
much progress with it; and of course it is a gamble that buffering
against the effect of temperature of facet-number will also protect
against the life-shortening effects of high temperature. [This
experiment was never completed satisfactorily. C.H.W.]

I always feel a bit disappointed that whenever you want to refer
to someone on developmental buffering and such subjects, you
always quote Schmalhausen and never my work. Admittedly we
were both developing very similar ideas at about the same time,
i.e. 1940, though at that time his were only published in Russian or
Ukrainian and little known over here. But I feel that mine were
more solidly based on facts—on a detailed study of 40 genes affecting
the *Drosophila* wing, in particular, as discussed in my *Organisers and
Genes* of that year—while in his case it was more a matter of theory
to bring genes into the story. (Again, at the same time, Paul Weiss
began to realize the importance of alternative developmental pathways, but didn't realize that they ought to be thought of in terms of
the genetic complexes which determine them.) And Schmalhausen
still seems to me, on re-reading him, to have got the theory muddled
up, not really distinguishing 'stabilizing selection type 1', which
holds the gene pool constant, and 'stabilizing selection type 2' which
builds up genotypes which determine developmental pathways
which exhibit what you call homeostasis and I call homeorhesis.
Finally, his distinction between 'adaptive modifications' and
'morphoses' seems to me a very artificial and almost metaphysical
one. I don't see that there is any reason to suppose there is any
essential physiological difference; what we presumably have is a

graded series of the effectiveness which natural selection has had during the past history in moulding the type of reaction to external stress. It is the *effectiveness* of selection, not its existence which is important. Just outside the climatic tolerance limit of a species, individuals of it must have been subjected to the external stress, but natural selection though operative has failed to turn a 'morphosis' into an 'adaptation'. On the other hand in my experiments with crossveinless and bithorax in *Drosophila*, the phenotypes were originally just what Schmalhausen would call morphoses, but selection could mould the genotype's reaction to the stress exactly as natural selection moulds the adaptive response in the case of salt acclimatization (also published in *Nature*—all the reprints of these papers are held up by a printing strike which is going on here).

I am just starting another experiment which may interest you. Selecting *Drosophila* for small body size, which usually reduces fitness; but I shall select simultaneously for good larval survival in competition with a tester stock. Then after some generations I shall hope to find that if selection is reversed (so as to aim at large size) the decline in fitness is greater then if one starts selecting for large size from the foundation stock. This is, of course designed to demonstrate a mechanism which makes it easier for selection to continue for long periods in the same direction, as in the lineages one used to call 'orthogenetic'.

I am looking forward to seeing you at this Chicago meeting in the autumn.

By the way, I was surprised a few months ago to get a letter from Gluschenko asking me to contribute an article on 'Darwinism and the Modern Genetics' to a Darwin centenary number of *Agrobiologia*. I don't know whether he was under the impression that 'genetic assimilation' is a Lysenkoist theory. Anyway I sent them the enclosed article. But I have had no acknowledgement of its receipt, and I suspect they will eventually say that it arrived too late for inclusion —they told me originally that the article would have to be in within a couple of weeks, so I had to write it very hurriedly. I thought that it was worth trying to get a fairly sensible account of modern theory into their journal for once.

Dobzhansky to Waddington, 15 August 1959

Many thanks for your letter and MS, which reached me here, in the wilds of Arizona, where I am doing some field work on *Drosophilae*.

First, concerning quoting Schmalhausen and not you. Having thought about it, I plead guilty and apologize. To be frank, I have also felt that the Edinburgh group is averse to quoting anybody's works but their own and their friends in Great Britain, but as a matter of fact this does not apply so much to you, and in any case I did not consciously try to redress the balance. And so, I promise to do better in the future.

I would like to defend Schmalhausen a little. Of course, he did not express things exactly as you did, but it seems to me that he had quite clear ideas concerning homeostasis, homeorhesis, etc., and concerning their roles in the grand scheme of evolution. I regard his *Factors of Evolution* as one of the 'basic books' establishing the biological theory of evolution, although it is written in a remarkably heavy language, both in the Russian original and especially in the English translation, which I have not 'edited' enough, for the good reason that editing might have meant in this case doing another translation job. Concerning modifications and morphoses, I don't see that Schmalhausen implied 'any essential physiological difference' between them. Sun tanning is a modification, but blistering is a morphose—the first is a reaction directed by natural selection in a channel which is adaptive, the latter is adaptively ambiguous, at least as it stands at present, but with further selection it might, conceivably, become a modification also.

Concerning Schmalhausen, let us also remember that he is one of the victims of Lysenko, and any support which we may give him, especially while he is still alive, is a good deed. You have done this and another good deed as well, in your 'Darwinism and modern genetics', which, I think, is excellent for the purpose. If it was written hastily, this, I assure you, is not obvious from reading it. You are probably right that they will not publish it, or publish with one of their 'prefaces', or, worst of all, publish only selected quotations which seem to be good to them. These people are completely dishonest, and one should be on one's guard with them—'Timeo Danaos et dona ferrentes'. [Actually the whole of my MS was published, without comment. C.H.W.]

Hope to see you in Chicago! Best regards to the family and to colleagues in your Institute!

11

CANALIZATION OF THE DEVELOPMENT OF QUANTITATIVE CHARACTERS

The characters of an adult organism are, of course, always the results of processes of development. In proposing theories of the genetical control of quantitative characters it has, however, been usual to start by considering the relations between the final character and particular (and mainly hypothetical) genes, attributing to them additive effects or more complex relations of dominance, epistasy, super-dominance and so on. The nature of the developmental system as a whole has in general been neglected, or brought in only at the end of the discussion as an afterthought to help clear up any difficulties left outstanding. I should like, as a student of development who cannot in any way claim to be an expert in quantitative inheritance, to try very shortly to approach the problem from the opposite angle and to examine whether our ideas about the general nature of development do not suggest certain ideas about the control of quantitative characters which may prove useful to those technically equipped to handle such matters.

Perhaps the most striking fact about development is that which has been referred to as canalization[1], that is to say the capacity to produce a particular definite end-result in spite of a certain variability both in the initial situation from which development starts and in the conditions met with during its course. The idea of canalization first arose in connection with the way in which developmental processes are integrated so as to produce a relatively small number of well-defined and distinct tissues and organs. A similar balancing-out of conflicting tendencies would seem, however, also to affect the development of quantitative characters. Thus for most animals there seems to be a 'normal' size, to which the adult individual often closely approximates even though it may have suffered various enhancements or retardations of growth during its lifetime as a consequence of factors such as the number of its littermates, the level of its feeding and so on. Such buffering against environmental influences is particularly characteristic of wild-type genotypes, and is often much less in individuals which are phenotypically abnormal owing to the presence of a strongly acting mutant gene (e.g. eyeless or aristopedia in *D. melanogaster*). These, as is well known, are in many cases markedly more variable than the wild type, and may show considerable sensitivity to agents such as temperature. There are many examples which demonstrate that the degree of canalization can be affected by selection. For instance, Landauer[2, cf. 3] has described the lower variability of races of polydactyl fowls which have been bred for this character for many years as contrasted with races into which the gene has been recently introduced. Further, he showed that races of fowls containing the mutants polydactyly, 'short upper beak', or Creeper, which had been selected for some generations for a low grade of expression of these characters, had become less sensitive to treatment with insulin (which can induce phenocopies of all these effects) than were normal races which had not been selected in this way. The selection in these cases must have built up genotypes which were to a considerable extent buffered against agents, either genetical or environmental, which tend to produce this group of developmental abnormalities. It is to a similar selection for the development of some optimum grade of expression, that we must attribute the relative constancy of the quantitative characters of the wild type.

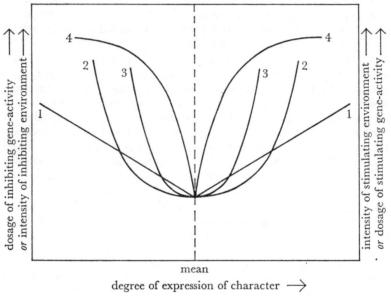

Figure 1.

Even the most well-canalized character is, of course, never entirely invariant. The canalization is expressed by the fact that extreme deviants from the mean are difficult to produce; the formation of lesser deviants may be stimulated much more easily. Within a certain range, in fact, the induced deviation from the mean may be linearly proportional to the applied environmental stimulus. One may envisage the situation in terms of a diagram such as figure 1. On the left side of this, the ordinates represent degrees of some environmental inhibition (e.g. lowered temperature) which leads to a deviation below the mean of the character in question, while on the right side the sense of the ordinates is reversed, so that they represent a condition (e.g. increased temperature) which tends to produce a deviation above the mean. In the most reactive type of system, we might expect the deviations to be linearly proportional to the stimuli (line 1), while more and more canalized systems are represented by lines 2 and 3. In both of these, there is a region of approximate linearity on either side of the mean, the situations being represented by lines which are convex towards the linear relationship shown in line 1. Such conditions might be termed 'smooth canalization'. One can also envisage the possibility indicated in line 4, which is concave toward the relationship of linearity. In such 'cusped canalization', abnormal environments would have little effect on the expression of the character until they attained a considerable degree of abnormality; threshold phenomena would be a special case of this general type.

The canalization of development protects the individual not only against fluctuations of the environment, but also buffers it to some extent against the effects of variations from the normal or 'wild-type' genotype. This effect is exemplified, in a rather extreme form, in the frequent occurrence of dominance, that is to say, situations in which a deviation from the normal in the form of a single dose of a recessive gene has no visible effect whatever. An abnormal gene affects the course of development (which is controlled by the genotype as a whole) by producing some alteration in the quantity or nature of the reacting substances. The mechanism of gene action is thus not essentially dissimilar to the means by which environmental effects on development are produced; and so if there is canalization against external influences, there must be a similar buffering against internal genetical abnormalities. This too could be represented by the diagram in figure 1. When considering genetic effects, we should have to take the ordinates as representing deviations from the normal genotype, measured in terms of the 'dosage' of the gene-activity we are considering. According to the simplest assumptions, we should think of the quantitative character under discussion as being controlled by numerous, perfectly additive genes, individually of small equal effects. In such a case, the ordinate scale would give the number of such mutant genes in the individual concerned, and gene-effect curves would be the rectilinear system 1. It is, in practice, not important to consider the situation in which the individual genes have effects of different magnitudes, since one can in any case not isolate them or measure such differences. The possibility to which I wish to draw attention is that the relation between the dosage of gene-activity and its effect on development may not be linear, but may be represented by one of the curves which exhibit canalization, such as 2, 3, or 4.

If we are dealing with a developmental system whose characteristics are those indicated by lines 2 or 3, there will be a certain range of variation around the mean within which the simple theory of control by additive genes would fit fairly well. The extent of this region would be greater in the conditions of line 2 and less for line 3; but in both cases the theory would break down entirely if applied to individuals which deviated greatly from the mean of the original population. In situations such as those represented in line 4, the theory of additive gene action would be quite inadequate even for close variations around the mean; dominance and epistasis could never be left out of account.

We have seen that there is evidence that, as might be expected *a priori*, the intensity of canalization can be increased by selection. Thus if a population is long selected for a certain optimum, one would expect it to pass from the situation of line 1 through that of line 2 toward line 3. In animals which have a well-developed wild type, which has been exposed to natural selection for long periods of time, one would expect that most quantitative characters would be canalized to a considerable extent, corresponding perhaps to line 3. If such a wild population were then to be artificially selected toward some other phenotype (e.g. towards larger size) it is to be expected that the hypothesis of additiveness of gene action would fairly soon prove inadequate. The phenotype would have been shifted, after a certain number of generations, into a region in which the curve relating to dosage of gene-activity to degree of developmental effect was far removed from simple linearity. Variation would then be dependant largely on dominance and epistatic effects; and the offspring of crosses between two similar deviants (two very large individuals, for example) might be expected to show a tendency to revert towards the original canalized mean.

In other cases, the population from which selection starts might have had a less intense canalization built up in it. It seems possible, for instance, that in many domestic animals, genotypic selection has not been intensive enough, or long enough continued, to have resulted in a strong canalization around some increased value of the desired characteristic. Indeed, if selection is carried on for some time in a number of different lines, and the lines then crossed, it would appear probable that a population would result in which the original canalization due to natural selection would have been largely broken down. In such a population the conditions might approximate to those of line 2. The region in which the assumption of linearity holds reasonably well would be larger, and this theory might prove an adequate guide for many generations of selection.

It would appear to be a matter of considerable practical importance to determine, for any population in which selection is to be practised, the characteristics of the canalization against which selection will have to contend. These characteristics cannot be adequately summarized merely by determining the deviations from the assumptions of linearity since, if the above view is correct, the magnitude of these deviations will depend on the extent to which the individuals concerned differ from the original mean.

If canalization is at all intense, as in situations corresponding to line 3 or, still more, line 4, selection will be difficult since much of the variability in gene-activity will be hidden by the developmental buffering. In such circumstances, it might be advantageous if some way could be found to break down the buffering system. If a powerful environmental influence shifted development somewhat out of its normal course, it is probable that the new course would be less strongly canalized, and the resulting adults more variable. This seems to be the case, for instance, for European cattle in the tropics. Selection of appropriate genotypes may be easier in such cases. Similarly, canalization is usually weakened in the presence of a mutant which causes considerable abnormalities in development. Even mutants which at first sight seem to have only rather unimportant effects (such as genes for coat colour or pattern) seem often to de-stabilize the general processes of development so that the adults are more variable in many different characters. It may be because of their effectiveness in disrupting canalization, which in the wild type conceals so much genetic variability, that genes for apparently inessential characters have been so frequently incorporated into man's selected breeds of domestic animals. And it may prove to be worth while, before attempting selection in a stock, to introduce into it a few genes which are not characteristic of the original wild populations. [1952]

12

EXPERIMENTS ON CANALIZING SELECTION

It is a platitude to point out that natural selection is not only responsible for the elaboration of evolutionary novelties, but that it has an even more universal function in the preservation of normality by the elimination of aberrant individuals. This second type of action has been referred to as stabilizing selection [1], and may operate in two rather different ways [2, 3]. In the first place, natural selection will tend to eliminate those alleles which in the normal environment cause the development of abnormal phenotypes; this may be called normalizing selection. A rather different type of natural selection will tend to remove those alleles which render the developing animal sensitive to the potentially disturbing effects of environmental stresses, and will build up genotypes which produce the optimum phenotype even under sub-optimal or unusual environmental situations. This type of selection has been referred to as 'canalizing selection', since it brings it about that the epigenetic systems of the animals in the population are canalized, in the sense that they have a more or less strong tendency to develop into adults of the favoured type.

It is well known to all geneticists that different strains differ in their sensitivity to environmental influences. A genetic variability on which canalizing selection could operate must therefore exist. There has, however, been rather little experimental work directly concerned with investigating the response to canalizing selection and the magnitude of the effects which can be produced, although many instances in which canalizing selection was actually practised could probably be found by a careful search of the older literature, particularly that of plant genetics.

Among the newer studies of the quantitative genetics of animals, the experiments of Mather [4] and Tebb and Thoday [5] perhaps come closest to this aim. They investigated the effects of selection and various other genetic procedures on the difference in the number of sternopleural bristles on the two sides of a *Drosophila*, and demonstrated that the genotype may have an effect on the magnitude of this difference. It may be presumed that the asymmetry between the two sides of the individual arises in response to slight differences in conditions between the two regions of the developing animal. Nothing, however, is known about the nature of these presumed differences during development. Reeve and Robertson [6] studied a situation which is perhaps of a similar kind; namely, the differences in the bristle numbers on adjacent abdominal segments in *Drosophila*. They have spoken of these differences as arising from 'intangible accidents of development'. It is, indeed, not entirely clear that either type of difference has any definite ascertainable cause. It is possible that some developmental processes have an inherently stochastic character, and the divergencies between the various elements of an essentially repetitive or symmetrical system may probably best be compared to 'noise', in the information-theory sense[3]. It is at any rate not at all clear that the genetic control of such a character can be taken as a guide to the nature of the genetic effects exerted on responsiveness to specific environmental stresses.

Falconer[7] and Falconer and Robertson[8] have also carried out experiments which involved selection of a type closely connected with canalizing selection. The former, working on *Drosophila*, selected those individuals whose bristle number on two abdominal segments lay closest to the mean of the population. Falconer and Robertson applied selection in mice for individuals whose weights were closest to the mean. In both these experiments the selection would have involved both the normalizing and the canalizing aspects of stabilizing selection, but in neither case was any definite environmental stress applied, and the relative importance of the two types of selection remains unknown. In the event, neither of the experiments revealed any marked response to the selective pressures applied, which, however, were rather small.

For these reasons it has appeared desirable to carry out an experiment designed to show whether canalizing selection can be actually effective in reducing the amount of phenotypic variation produced by a definite and known environmental stress.

MATERIALS AND METHODS

All the experiments were made on *Drosophila melanogaster*. The development of normal wild-type individuals of most stocks of this species is already rather highly canalized, so that the easily applied environmental stresses produce rather little phenotypic divergence from normality. It might therefore be rather difficult for a selection experiment, lasting a comparatively few generations, to produce marked changes in canalization, and in particular increases of it. The experiments were therefore made on mutant stocks in which, as is well known, the phenotype is often much more variable and sensitive to environmental influences than is that of the wild type. Four mutant types have been studied: Bar, dumpy, cubitus-interruptus, aristopedia-Bridges.

1. *Bar*. The number of facets in the eye is markedly reduced, the effect being stronger in the homozygous females than in the hemizygous males. The number of facets produced by a given genotype is known to be greater at a low temperature than at a higher one. The facet numbers can be counted only with some difficulty in the living fly, and estimations of average facet numbers in a population were therefore made by counts on eyes prepared after death. The parents for carrying on the selection lines were selected on the basis

of estimated eye sizes, as judged by inspection with a dissecting binocular.

2. *Dumpy*. This gene produces a shortening of the wing which is more extreme at high temperatures than at low ones. The effect was estimated by measuring the wing length from the base of the wing to the most extreme point of the rounded tip.

3. *Cubitus-interruptus*. The gaps which this gene produces in the posterior longitudinal vein are more extreme at low than at high temperatures. They were estimated in terms of an arbitrary series of grades founded on that used by Stern and Schaeffer[9] (see Waddington, Graber and Woolf[10]). In the figures given in this paper the numerical indices attached to the grades are one less than those used in the paper by Waddington, Graber and Woolf; that is to say, the wild type scores as 0 and the most extreme type as 6. These grades are of course quite arbitrary and can be used only for comparative purposes. The stock employed was a multiple recessive stock containing $ci\ gvl\ ey^R\ sv^n$.

4. *Aristopedia-Bridges*. This is a fairly weak allele of the well-known gene which tends to convert the aristae into leg-like appendages. The strain used was extracted from the F_2 of a cross between the aristopedia stock maintained in the laboratory and an Oregon-R wild type. The effect is more extreme at low than at high temperatures. It was estimated by a series of arbitrary grades, similar to that employed by Waddington and Clayton[11]. Among the more extreme modes of expression there are several in which the size of the aristal organ becomes reduced. In assigning numerical values to the grades these have been given relatively low values, similar to that of phenotypes in which only part of the arista has been converted into a leg-like structure. It is, however, by no means certain that this is the correct way of envisaging the situation. One of the effects of aristopedia is to cause reductions in the leg, and it may be that these reduced aristal organs really represent an extreme grade in the expression of the genic action, in which the arista is first converted into a leg and then reduced. If that were the case, the reduced organ should perhaps have a higher numerical index rather than a lower one corresponding to its size. It is unfortunate that, as it turned out, a considerable proportion of the populations kept at low temperatures had phenotypes within this range. As will be seen later, the selection practised had little effect in altering the average phenotype of these cold-temperature populations, and it may well be that owing to the uncertainties of the grading system very little selection was in fact exerted on them.

Three types of experiments have been carried out. The first two were both varieties of what may be called alternate selection. In these the strain was kept for one or more generations at one temperature, then moved to the other temperature for one or more generations, and then back again to the first set of conditions, and so on alternately. Breeding always involved selecting against the known environmental effect of the temperature. That is to say, if, as in Bar, a low temperature tends to cause an increase in the number of eye facets and a higher temperature a reduction in facet number, one would select from a population reared at a low temperature those with the smallest facet numbers, and from a population bred at a high temperature one would select those with the largest number of facets. The two experiments carried out according to this plan differed only in that, in the first of them, one generation at a low temperature of 18°C alternated with two generations at the high temperature of 25°C, while in the second experiment only one generation at the high temperature followed one generation at the low. In a variant of the first type of experiment two generations at 25° were run simultaneously with one at 18°, and individuals from the second 25° generation were then crossed with those from the 18° generation to produce two new selected lines, one of which was again given two generations at 25° while the other had one generation at 18°, when the process was repeated.

A very different and, as it turned out, much more effective type of selection was applied by means of family selection. In this, a series of pair-matings were set up, and from each pair of parents a number of offspring were reared at 25° and a number at 18°. Family averages of the difference between the phenotypes at these two temperatures were then ascertained by measuring a certain number (usually twenty) of the offspring at each temperature. Selection was made on the basis of these family averages, lines being carried on by breeding together sibs from the families which exhibited the least environmental sensitivity and in which the difference between the offspring at the two temperatures was smallest.

RESULTS

1. *Alternate selection*

The results of the experiments carried out with Bar, dumpy and cubitus-interruptus are given in tables 1 to 3. It will be seen that there was considerable progress in reducing the environmental sensitivity of the Bar stocks. In dumpy and cubitus-interruptus only rather slight progress was made. This result may be partly due to the necessity to use, in these two stocks, a somewhat inadequate

Table 1. Bar. Mean facet number per eye (two sides averaged)

Generations	1	2	3	4	5	6	7	8	9	10	11	12	13	14	15	16	17	18
Temperature	25	25	18	25	25	18	25	25	18	25	25	18	25	25	18	25	25	18
Male	91	82·6	222·8	100·0	117·7	191·6	91·0	101·0	182·3	90·0	127·9	183·8	116·0	106·2	196·0	122·8	110·8	204·8
Difference, 18°—25°		140·2	122·8		73·9	100·6		81·3	92·3		55·7	67·8		89·8	73·2		94·0	
Female	46·9	51·8	160·1	66·8	56·7	107·3	61·0	69·3	95·5	58·4	73·7	102·5	68·0	63·3	103·4	71·0	71·5	102·7
Difference, 18°—25°		108·3	99·3		50·6	46·3		26·2	37·1		28·8	34·5		40·1	32·4		31·2	

Table 2. Dumpy. Wing lengths (two sides averaged)

	Generation	1	2	3	4	5	6	7	8	9	10	11	12
	Temperature	25	25	18	25	25	18	25	25	18			
	Wing length male	55·8	53·4	70·4	53·3	56·5	76·35	57·65	61	78·4			
Expt. 1 {	Difference, 18°—25°		17·0	15·1		19·85	18·7		17·4				
	Wing length male	58·8	56·95	75·65	59·9	63·3	76·8	62·1	63·1	80·1			
	Difference, 18°—25°		18·7	15·75		14·7	13·5		17·0				

Expt. 2

Temperature	25	18	25	18	25	18	25
Wing length male	52·8	75·35	58·5	76·15	58·6	77·65	61·45
Difference, 18°−25°	22·55	16·85	17·65	17·55	19·05	16·2	
Wing length female	55·05	77·45	61·75	74·45	62·7	78·6	63·95
Difference, 18°−25°	22·4	15·7	12·7	11·75	15·9	14·65	

Expt. 3

Temperature	25	25	18	25	25	18	25	25	18	25	25	18
Wing length male	58·1	57·8	76·45	56·1	60·15	75·45	60·2	60·1	76·25	61·5	63·1	79·35
Difference, 18°−25°		18·65	20·35		15·3	15·25		16·15	14·75		16·25	
Wing length female	59·48	59·4	77·45	60·15	60·85	74·45	59·75	64·0	79·55	64·1	63·5	79·5
Difference, 18°−25°		18·05	17·3		13·6	14·7		15·55	15·45		16·0	

Table 3. Cubitus-interruptus. Mean grade (two sides averaged)

Generations	1	2	3	4	5	6	7	8	9	10	11	12	13	14	15
Temperature	25	25	18	25	25	18	25	25	18	25	25	18	25	25	18
Male	138	127	202	151	150	212	140	134	208	140	159	260	156	140	196
Difference, 18°−25°		75	51		62	72		74	68		51	54		56	
Female	93	96	182	98	116	199	127	83	170	113	132	151	133	130	194
Difference, 18°−25°		86	84		83	72		87	57		19	18		64	

system of scoring by arbitrary grades. The imperfections of this system probably reduce the amount of selection that can be exerted, and it probably reflects only very inaccurately the actual situation in the stocks which eventually arise. In spite of this, some slight reduction in environmental sensitivity has been achieved in nearly all cases. The magnitude of the effects, however, is so slight that it could hardly be taken to be sufficient to establish the reality of the process.

2. *Family selection*

Family selection experiments were carried out with Bar and with aristopedia-Bridges. The results are shown in tables 4, 5 and 6. In

Table 4. Bar. Facet number in females. Pedigree of family selection experiment. The figures give the differences between the means of the family-samples reared at 25° and at 18°. The plain figures give the average of the differences of all families tested; the figures in brackets the difference in the family selected to carry on the line.

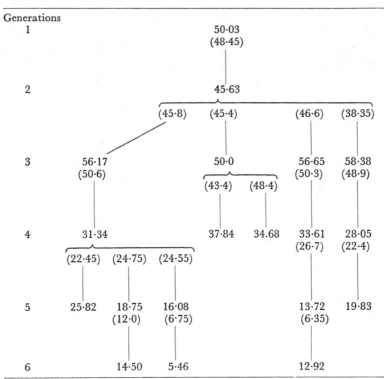

both cases very marked reductions in environmental sensitivity have been achieved.

In these experiments selection was carried out on the basis of the average magnitude of the difference between the high- and low-temperature sibling groups. No attention was paid to the absolute magnitudes of the phenotypic indices. In the case of Bar this has resulted in the production of a relatively temperature-insensitive eye of a size intermediate between that which was characteristic of the low- and high-temperature populations at the beginning of the experiment. Some representative figures on this point are given in table 5. With aristopedia the result was rather different (table 6).

Table 5. Bar. Facet numbers in females in unselected controls, and in three families resulting from selection.

	No. of Families	Average family mean, at 18°C, and s.e.	Average family mean, at 25°C, and s.e.	Difference, family means, 18°C and 25°C
Unselected	9	156·25±4·81	55·46±0·81	100·80±5·42
Selected 6*a*	8	106·02±1·17	91·51±1·67	14·50±2·29
Selected 6*b*	7	100·76±1·38	95·30±1·70	5·46±1·98
Selected 6*c*	8	111·39±1·72	98·48±2·67	12·92±3·05

Table 6. Aristopedia-Bridges. Family mean grades.

Generation	No.	18°	25°	Difference
0	10	14·01±0·085	7·16±0·215	6·84±0·177
1	20	13·36±0·130	7·46±0·152	5·89±0·157
2	20	13·74±0·067	7·89±0·141	5·90±0·167
3	20	13·58±0·062	7·39±0·184	6·20±0·20
4	20	13·36±0·07	8·72±0·13	4·65±0·14
5	20	13·60±0·071	10·50±0·179	3·10±0·190
6	19	13·44±0·626	11·13±0·286	2·31±0·268

The reduction in temperature sensitivity has in essence been brought about by increasing the abnormality of the high-temperature population, in which initially the gene was rather weakly expressed. There has been scarcely any accompanying reduction in the expression in the low-temperature group. As was pointed out in the section on methods, this is probably a consequence of the imperfections of the scoring system, which seems likely to be particularly unreliable in discriminating between the various higher grades of expression, so that it seems probable that in fact very little selection of any kind was actually exerted on the low-temperature populations.

It may appear from table 4 that the response to selection has been greater then the selection-pressure applied, since in many instances the later generations show a smaller average family difference than that found in the family which was selected in the immediately preceding generation. However, the figures given in brackets in that table, which record the difference between the 25° and 18° groups of the families selected, do not tell the whole story, since selection was also practised *within* the families, by selecting individuals which showed little effect of the temperature treatment. It is therefore impossible to give any accurate analysis of the rate of response to the selection-pressure. The bracketed figures are given merely to show that the response cannot be considered a slow one.

In view of the interest which has been taken in asymmetry as an indication of developmental stability, it seemed interesting to inquire whether successful canalization against the effect of temperature was accompanied by a greater stability against the unknown environmental variations to which asymmetry may hypothetically be attributed. The asymmetry of facet numbers was therefore estimated, the measure used being the difference in the logarithms (to base 10) between the larger and the smaller of the two eyes. As can be seen from table 7, there was little alteration in the asymmetry during the course of selection; in the twenty-three families measured after selection, the asymmetry at both temperatures was slightly less than it had been in the set of pair-matings from which the selection started, but the difference is small and non-significant. When one compares the families at the end of selection with the particular foundation pair-mating from which they were all derived, the asymmetry at 25° is found actually to have increased. One may conclude that selection for better buffering against the external temperature has had little or no effect on asymmetry.

Table 7. Mean asymmetry in facet number ($\log_e$ scale).

Selected for	Progeny of	18°C	25°C
0 generations	20 pairs	$0·102 \pm 0·025$	$0·095 \pm 0·018$
0 generations	Foundation pair	$0·115$	$0·069$
6 generations	23 pairs	$0·093 \pm 0·026$	$0·081 \pm 0·026$

DISCUSSION

The family selection experiments, particularly that with Bar, provide a very clear example of the effectiveness of canalizing selection. In the unselected population the eyes of individuals reared at low temperatures have about three times as many facets as those reared at 25°, while by the end of the selection experiments the difference has been reduced to about ten per cent.

This reduction in environmental sensitivity has occurred in an epigenetic system which is abnormal in the sense that it is affected by an unusual allele, namely, the dominant mutant gene Bar. There seems no reason to doubt, however, that similar processes of selection would be effective in reducing the environmental sensitivity of a population in which no obviously abnormal allele was present. The array of genotypes which fall within the broad category of the wild type all give rise to epigenetic systems which are, in general, rather insensitive to external stresses, but it seems inevitable that amongst them there must be some genotypes which provide better developmental buffering than others. It was not by any means obvious, however, that such an abnormal developmental system as that resulting from the Bar mutant could be brought to a condition of such temperature-insensitivity as we have found. Since canalizing selection, operating through families, has been so extremely effective on this abnormal developmental system after only five generations, one is left with a feeling of considerable confidence as to its powers when acting on more normal populations.

The comparative failure of systems of alternate selection is perhaps not very surprising. When in an 18° population of Bar one selects the individuals with the smallest eyes as parents of the next generation, one is, of course, not only exerting canalizing selection in favour of genotypes which are insensitive to the temperature effect, but one is also applying progressive selection in favour of genotypes which would tend to produce a small number of facets under any environmental circumstances. In the next generation, reared at 25°, selection for large eyes would again favour well-canalized genotypes but would now also involve a progressive selection of the opposite kind to that in the previous generation, since it would favour genotypes which tended to produce large facet numbers under all circumstances. The canalizing selection would only be effective in so far as the two opposite types of progressive selection cancelled each other out, and in proportion to the importance of canalizing factors as opposed to facet-increasing or facet-reducing factors in determining which individuals will be selected. There is no way of telling *a priori* which will be the more important elements in the situation. The fact that in these experiments there was in several cases an immediate slight increase in canalization, but that the rate of progress was not maintained thereafter, suggests that the presence of alleles of different quantitative effectiveness has made it very difficult by this procedure to concentrate alleles whose effect is primarily on environmental sensitivity.

The mathematical theory for selection for canalizing genes has never been worked out. It is, however, easy to show that selection which operates only against the phenotypic effect of a single environment, such as that which was exerted in the alternate selection experiments, would often tend to concentrate quantitatively-acting genes rather than canalizing genes. Consider, for instance, a population which contained a relatively rare allele a whose effect in all environments is to reduce the size of a given phenotypic measure. Let the frequency of a be p. Further, suppose the population contains an allele b with a frequency of P and with the effect that the homozygote bb is less environmentally sensitive. Now, suppose the population to be in an environment which tends to increase the phenotypic measure, and let selection be applied against this environmental effect. We will suppose that the selected group includes all the genotypes containing aa, and that the reduction in phenotypic sensitivity brought about by b is such that the genotypes $Aabb$ would fall within it. Then in the next generation

$$p_1 = \frac{2p^2+2pqP^2}{2p^2+4pqP^2} = \frac{p+qP^2}{p+2qP^2} \quad \text{and}$$

$$P_1 = \frac{2p^2(P+PQ)+4pqP^2}{2p^2+4pqP^2} = \frac{pP+2qP^2}{p+2qP^2}.$$

Thus, for the proportional rates of increase of a and b we shall have

$$\frac{p_1-p}{p} = \frac{q[p+P^2(1-2p)]}{p(p+2qP)} \quad \text{and} \quad \frac{P_1-P}{P} = \frac{2qP(1-P)}{p2qP}.$$

The 'quantitative gene' a will increase proportionately faster than b if

$$\frac{p+P^2(1-2p)}{p} > 2P(1-P),$$

i.e. $P^2 > p(P-Q)$.

It is clear that, with appropriate values of the frequencies, either the quantitatively-acting or the canalizing genes will increase the more rapidly. Without some prior knowledge of the relative frequencies of genes of these two kinds, one cannot predict what the outcome will be.

The lack of response of asymmetry to selection which was effective in reducing sensitivity to the particular environmental variable of temperature is worth noting. But it should occasion no surprise.

Reasons have been offered previously [3] for expecting differences between comparable regions within a single individual body to be controlled by genes which are not the same as those which affect the reaction of the body as a whole to the external environment.

SUMMARY

1. Mutant stocks whose phenotype is affected by temperature were kept at 18° and 25°C, and selection was applied in an attempt to reduce the magnitude of the phenotypic difference caused by the temperatures. The stocks used were Bar, dumpy, cubitus-interruptus and aristopedia-Bridges.

2. When the stocks were kept for one or more generations at one temperature, then transferred for the next generation or two to the other, and so alternately, selection applied against the phenotypic effect of the temperature was only slightly effective in reducing the differences between the high- and low-temperature phenotypes. It is suggested that this was due mainly to the fact that such selection would be expected to operate on quantitatively-acting genes as well as on genes controlling developmental buffering.

3. With family selection, in which the offspring of a pair-mating was divided into two lots, one kept at each temperature, and selection was made on the basis of the differences in family means, progress was rapid. In the unselected Bar stock, the facet number at 18° was initially about three times that at 25°; after six generations of family selection the difference had been reduced to about ten per cent, the phenotypes at both temperatures being about intermediate between those seen at the beginning of the experiment. With aristopedia the reduction in the difference in family means was also striking, but in this case was achieved by increasing the abnormality of the 25° phenotype, and hardly at all by lowering that of the 18° population. It is pointed out that this is probably the result of an inadequate scoring system.

4. Although the selection in Bar stocks was effective in increasing developmental canalization against external environmental changes (temperature), it had little or no effect on the asymmetry of the facet numbers. [1960]

13

SELECTION FOR DEVELOPMENTAL CANALIZATION
(WITH E.ROBERTSON)

It was shown some years ago[1] that in stocks of *Drosophila melano-gaster* containing the mutant Bar the effect of temperature on eye size is under genetic control. In normal laboratory stocks the eyes are larger at 18°C than at 25°C, but it was possible by selection to produce a line in which the difference at these two temperatures had been reduced to a small fraction of what it is in the unselected founder stock. However, noteworthy success in improving the canalization of Bar against the effect of temperature was only achieved by the somewhat artificial method of family selection, which must also have involved a considerable amount of inbreeding. It was therefore decided to carry out another experiment in which selection would be applied both to improve and to reduce canaliza-tion, using other breeding procedures. The experiments also give some information about the effects of submitting a population to two different selection processes simultaneously ('disruptive selection').

PLAN OF THE EXPERIMENTS

A base population was set up as follows: from a laboratory culture of Bar ten new bottles were set up by merely shaking over flies from the stock bottle, these parents being removed after a few days. When the new generation emerged, from each of these ten bottles six small-eyed males were taken and mated with six large-eyed females and separately six large-eyed males were mated with six small-eyed females. After allowing 24 hours for mating the two groups originat-ing from each bottle were placed together and allowed to lay. This gave rise to ten stocks, each of which should have contained a large fraction of the total genetic variation available in the initial popula-tion. The later selection was carried out in the same manner in each of these ten replicate stocks.

To ascertain the eye size in the foundation generation of the stocks, 400 larvae were collected from each stock; 200 of these were allowed to develop at 25°C while the other 200 were kept for 48 hours at 25°C and then transferred to 18°C until they had completed their development. From the flies emerging from these two temperature treatments a random sample of twenty males and twenty females were used for facet counts. The flies were digested in 10 per cent NaOH, the corneas then removed, stained in 1 per cent pyrogallic acid, and squashed under a coverslip, tearing if necessary to ensure flattening, and the facet numbers counted under a compound microscope. The facets on both eyes were counted, but only the means of both sides will be quoted here since no additional information emerges from consideration of the sides separately.

From each of the ten replicate founder stocks, two selected lines were set up: a 'canalizing' and an 'anti-canalizing' line. Canalizing lines were set up as follows: from the part of the stock treated with low temperature (which should enlarge the eye) the five males and five females with the smallest eyes were selected from a sample of 100. This selection was not based on precise facet counts but on the judgement of eye size, flies being inspected with a binocular microscope at a magnification of 15 ×. From the part of the stock cultured throughout at 25°C, the five males and five females with the largest eyes in a sample of 100 individuals were selected. These ten males and ten females were put together and allowed to mate at random. From the eggs produced by this mating, two batches, each of 150 larvae, were again subjected to the two temperature treatments, and when the adults emerged the selection procedure was repeated.

The selection procedure for anti-canalization was essentially similar except that individuals with the largest eyes were selected from the low-temperature treatment and those with the smallest eyes selected from the high-temperature treatment. In all these lines selection was practised on the right eye only.

Precise counts of facet numbers were made at the fourth and ninth generations of selection. Throughout the work, the 18°C temperature fluctuated within a range of ± 1°C. During most of the selection the 25°C temperature fluctuated within a range of ± 2°C but in the fourth and ninth generations the flies to be used for facet counts after 25°C treatment were grown in an incubator in which the temperature was controlled to less than half a degree.

With this breeding procedure it will be apparent that the canalization and anti-canalization lines were kept separate from the beginning of the experiment with no gene flow between them. However, within each of these lines there was no attempt to control the mating between the parts of the stock which had been cultivated at high and at low temperatures, the selected flies drawn from these two treatments being mixed and allowed to mate at random.

Within each of these two lines, some individuals were selected for large eyes, others for small eyes. This amounts to the practice of 'disruptive selection'. The form this has taken in the present experiments is rather different from the way disruptive selection has been carried out in other experiments on the subject, such as those of

Thoday and his collaborators [2], since in the present work selection for two optima was performed on a population which was also subjected to controlled environmental variations which affect the phenotypes under selection. We will return to this point in the discussion; in the meantime it may be pointed out that selection of a population for two different optima would be expected, on the face of it, to lead to an increase in phenotypic variance.

RESULTS

The effectiveness of selection for and against canalization

Table 1 shows the mean numbers of facets in the eyes of flies in the foundation population and in the canalization and anti-canalization lines at generations 4 and 9, under the two temperature treatments, 18°C and 25°C. We should expect the canalizing selection to reduce the difference in facet number between the two temperature treatments and anti-canalizing selection to increase it. It is clear that these expectations are, in general, fulfilled, although in males of the anti-canalizing line the increase does not progress steadily, but is larger in F_4 than in F_9.

Table 1. Mean number of facets per eye (right eyes only, twenty eyes counted per replicate, averages of ten replicates).

	18°C	25°C	Range 18–25°C	Difference in range	t
Males					
Foundation stock	163·9	98·5	65·4		
F_4					
Anti-canalization lines	185·8	97·4	88·4⎫	29·3	5·45**
Canalization lines	175·7	116·6	59·1⎭		
F_9					
Anti-canalization lines	164·1	88·7	75·4⎫	20·6	2·62*
Canalization lines	197·4	142·6	54·8⎭		
Females					
Foundation Stock	137·8	83·1	54·7		
F_4					
Anti-canalization lines	142·4	87·1	55·8⎫	11·4	2·75*
Canalization lines	145·3	100·9	44·4⎭		
F_9					
Anti-canalization lines	133·8	71·9	61·9⎫	21·4	3·88**
Canalization lines	157·5	117·0	40·5⎭		

* Significant at $P=0.05$ ** Significant at $P=0.01$

In the canalization line selection was for small eyes at 18°C and for large eyes at 25°C. The results show that the size of the eye at 25°C has actually increased, but it also increased, though to a lesser extent at 18°C. Thus what has occurred is a general improvement throughout the whole population of one of the two favoured phenotypes.

In the anti-canalization line the results were more complex. Selection was for increased size at 18°C and in the first four generations this was effective. However, between generations 4 and 9, the response was reversed and at the end of the experiment the eye size was very slightly less than it had been in the foundation stocks at 18°C. The selection for small eyes at 25°C had, in the meantime, made slow progress with a slight fluctuation upwards at generation 4 in the females. Thus in the early generations there was an improvement in the adaptation at 18°C with little overall change (averaging males and females) at 25°C, but at the end of the experiment there had been a slight improvement in the 'adaptation' to 25°C accompanied by a slight deterioration of it for 18°C.

The effect of the disruptive selection on the phenotypic variance of the various lines was investigated by calculating the coefficients of variation of the various populations. The data are given in table 2. The main body of the table gives the coefficients of variation within the measured individuals (twenty males and twenty females) in the ten families at the original or parental generation (P) and at F_4 and F_9 for both the canalization and anti-canalization lines. The lower two horizontal panels of the table give the means of these coefficients, and 'standard deviation factors' suitable for assessing the significance of differences between the means. These two items have been calculated as follows:

'*Means*'. The standard error of a coefficient of variation V based on N readings is approximately $V/(2N)^{1/2}$. That is to say, the standard error is proportional to the coefficient. A simple arithmetical average of the coefficients would therefore be biased and misleading. A transformation to log V will however approximately stabilize the standard errors for cases such as this, where N remains constant. The 'means' given have therefore been calculated by taking the antilog of the mean of the logs of the coefficients for the ten families in each line.

'*Standard deviation factors*'. The standard deviation of the logs of the coefficients was calculated, and then converted to the antilog. The confidence limits of the 'means' can be found by multiplying and dividing the 'mean' by the appropriate multiple of the 'standard deviation factors' given in the table.

It will be seen from the upper part of the table that in all cases the mean coefficients of variation in the selected lines (of both kinds) had increased over that in the foundation stocks. The increases were considerably greater in the anti-canalization line than in the other. In both lines most of the increase had occurred already at generation 4, but, as the s.d. factors show, there is between generations 4 and 9 a continued increase in dispersion of the means. This is brought about mainly by a further increase in coefficient of variability in a few families, e.g. the canalization line no. 4, the anti-canalization nos. 6 and 8.

Table 2. Coefficients of variation. (In each group, the upper figures under F_4 and F_9 apply to the canalization line, the lower two figures to the anti-canalization line).

	18°C						25°C					
	♂			♀			♂			♀		
Family	P	F_4	F_9	P	F_4	F_9	P	F_4	F_9	P	F_4	F_9
1	18·5	11·6	10·7	14·9	17·5	10·8	14·9	17·7	18·1	12·7	14·0	14·9
		18·4	14·7		20·6	15·9		21·7	33·3		19·6	25·5
2	12·0	21·3	13·4	19·5	20·7	15·6	14·3	26·6	18·7	11·9	22·9	15·7
		13·6	14·8		20·1	15·7		18·5	23·6		27·1	23·5
3	6·7	16·1	12·8	11·7	13·9	12·0	11·1	17·7	13·1	14·0	18·9	12·6
		20·4	16·6		19·8	22·2		14·3	19·3		27·8	32·3
4	14·3	13·7	38·6	10·2	20·9	45·4	18·2	14·6	17·6	19·7	26·3	42·1
		21·7	20·1		29·8	22·9		37·7	19·1		23·6	23·5
5	13·3	12·4	16·1	16·3	18·1	21·9	16·0	15·4	29·9	16·4	20·0	22·1
		16·0	21·6		21·4	18·6		31·0	25·3		25·1	14·2
6	11·0	15·3	17·0	10·0	23·4	15·4	15·4	15·9	15·9	11·2	20·3	19·1
		11·4	30·9		20·6	28·9		16·9	38·6		19·5	34·9
7	14·2	15·2	26·3	15·4	28·3	24·7	14·5	19·6	23·3	18·4	15·7	21·3
		17·5	26·6		22·1	28·9		21·6	28·2		16·3	25·6
8	17·7	20·8	27·0	18·1	20·1	13·4	22·8	24·6	22·0	11·1	21·6	14·5
		14·4	54·0		47·7	37·9		35·9	48·2		42·6	50·2
9	13·2	18·5	13·6	17·9	15·0	14·5	21·7	17·1	15·7	18·7	18·7	22·0
		19·1	17·9		28·7	12·5		30·9	22·9		26·9	23·3
10	13·5	20·6	13·2	15·6	22·7	16·9	14·?	12·7	18·7	14·8	18·1	13·7
		13·2	36·6		23·4	33·1		35·9	29·2		31·7	25·3
MEANS	13·0	16·2	16·3	14·4	19·5	17·4	15·8	17·9	20·3	14·4	19·4	18·6
		17·3	23·4		24·4	22·3		23·2	27·3		25·1	26·5
S.D.	1·33	1·24	1·47	1·27	1·26	1·52	1·26	1·25	1·37	1·26	1·20	1·43
FACTORS		1·30	1·53		1·32	1·44		1·40	1·34		1·31	1·39

DISCUSSION

The results of the main experiment on canalization confirm those of the previous work; the sensitivity of the Bar stock to temperature differences is under genetic control and selection for greater sensitivity (against canalization) or lesser sensitivity (for canalization) is quite successful. This is so also in conditions which involve random mating and individual selection, as they did here, instead of the family selection which was practised in the previous work.

The selection for increased canalization not only reduced the sensitivity of the line to temperature differences but also brought about a general increase in facet number. This is the character which was selected for in the high temperature and against in the low temperature. Selection of this kind must be expected not only to increase the frequency of genes which reduce temperature sensitivity but also to bring about changes in the frequency of genes which tend either to increase or decrease eye size at all temperatures. The relative importance of these various effects in any particular course of selection will depend on the distribution of the various kinds of genetic variation available in the base population. The results of these experiments on canalizing selection lead to the conclusion that the founder population in this case contained a good deal of easily utilizable genetic variation tending to increase eye size at all temperatures. In the anti-canalization experiments, in which there was selection for increased eye size at low temperature and for reduced size at high temperature, there was an increase in eye size in the first four generations for males and females at low temperature and for females at high temperature, while the males at high temperature declined very slightly in facet number. These changes were probably produced by utilizing the same genetic variation as played such a large part in the canalization experiments. In the later generations of the anti-canalization line, the effect was reversed and there was some decline in eye size throughout the whole population.

In both the canalization and anti-canalization lines, in fact, the final result at the end of the experiment was an improvement in the character selected for at 25°C, whether this was increased size (canalization line) or decreased size (anti-canalization line). This presumably indicates either that heritabilities of eye sizes in general are higher at 25°C than at 18°C; or, perhaps more plausibly, that rearing at the higher temperatures gives the flies some reproductive advantage in a panmictic population, perhaps in readiness of males to mate, or in the number of eggs laid by females.

Before discussing the results of the 'disruptive selection' that operated in the two lines, it will be as well to consider in general terms the importance of this type of selection. As Thoday has often pointed out (e.g. [2], 1964) laboratory experiments on disruptive selection owe their main interest to the light they may throw on the conditions under which sympatric species-divergence may occur in nature. In a natural population, disruptive selection would occur if the habitat, which will usually be to some extent heterogeneous, contains two or more different environmental niches which differ from one another in selection optima. Individuals inhabiting one niche will then be selected by a criterion which is not the same as that applying to individuals in the other niche. The evolutionary consequences of these disruptive selection pressures will be affected by a number of other factors. In the first place, the rate of flow of genes between the sub-populations in the two niches will depend on the degree to which successive generations of a sub-population return to the same niche as that inhabited by their parents, that is to say, on the development of continuing habitat-preferences within the sub-populations. Again, the gene flow will be affected if there is any development of mating preferences, which lead members of a sub-population to mate with each other more frequently than with members of the other sub-population. A final factor, with which the present experiments are particularly concerned, is the possibility that the niches exert differential effects on the developing phenotypes of a kind which is relevant to the selection optima. Since the niches with which we are concerned are different enough to exert different selection pressures, and since most organisms notoriously show some capacity for phenotypic adaptation, this last factor would be expected to be of importance in most natural situations.

The largest body of experimental work on disruptive selection is that which has been carried out for some years by Thoday and his collaborators [2]. The plan of the earlier experiments corresponded to a natural situation in which habitat-preferences were more or less strongly developed. Separate lines were kept (e.g. a 'high' line and a 'low' line for bristle number) with different selection pressures exerted on them, and various controlled degrees of gene flow allowed between them. In these circumstances, it was usually found that the lines diverged. This would seem not unexpected, although Thoday expresses some surprise at the result. It is, however, obvious that divergence would occur if there was no gene flow between the lines; whether it occurs at any given rate of gene exchange will depend on the values of various genetic parameters in the population, such as the heritabilities of the characters, selection intensities, dominance and epistatic effects etc. The basic theory of the situation does not appear to have been worked out, nor have these parameters been estimated. However, it is not clear that such situations can have much relevance to the possibility of sympatric divergence in nature, since they depend on the existence of some previously developed system of habitat-preferences, without which the whole population could not be regarded as separated into two sub-populations between which there is gene exchange. The relevance of laboratory experiments on disruptive selection to situations in nature has also been discussed by Maynard Smith[3] with particular reference to the stability or instability of any polymorphism produced.

In a recent experiment, which has been rather briefly reported, Thoday and Gibson[4] (Thoday[5, 6]) have dispensed with anything corresponding to habitat preferences. In each generation, flies selected either for high or for low bristle number were all placed together and allowed to mate panmictically; and from their progeny, selection was again made for high or low bristle number, with no reference to the character of the parents. This experimental set-up is comparable, in its use of panmixis and of purely individual selection unrelated to the parental characters, to that of the present experiments; and both experiments would seem to provide a good model of a natural population faced with two environmental niches in relation to which no habitat-preferences have been developed.

The results of the two sets of experiments have, however, been somewhat different. In the work reported here there has been rather slight increase in the phenotypic variance exhibited by populations from the canalization lines when kept in a constant environment, while in the anti-canalization lines the increase was only about twice as great. Moreover, the selection has not led to a simultaneous increase in adaptedness to the two selection criteria applied. In the canalization line, the population as a whole became better adapted to one of the niches, i.e. the 25°C environment in which selection operated for large eyes, and worse adapted to the other. In the anti-canalization line, the overall result was again a slight increase in adaptedness to 25°C, in which, in this case, selection was towards smaller eyes.

Thoday and Gibson's experiments, on the other hand, led to a great increase in phenotypic variance, brought about by progress in both directions towards the two selection optima of high bristle number or low bristle number. The difference between the results of the two experiments could, perhaps, find its explanation in differences in the rate of response to selection in the two directions in the early generations. In each generation, there will be a panmictic population of parents, equal numbers of which have been selected for criterion A and for criterion B. Matings will occur of types AA, AB and BB. The first of these will be expected to give offspring which approach more nearly the selection criterion A, and the last more offspring fulfilling criterion B. As soon as any noticeable progress is made towards either selection-optimum, it will clearly be reinforced in later generations, especially if the selection pressure is high enough to ensure that nearly all the individuals selected for criterion A in any generation come from parents who were also selected for A. Thoday and Gibson's selection pressure was fairly severe; in each generation they selected eight high females from among a total of eighty females, in which forty were the progeny of eight parental high females which had had an equal chance of mating with high or low males. Even with completely random mating, one would expect four of these parental females to have mated with high males, so that the chances that the selected high offspring would come from a HH mating was very considerable. In so far as this occurred—and supposing a similar situation to hold for the selected low individuals —the two selection lines would in effect be quite distinct from one another even though an opportunity for cross-mating had been provided. Their divergence would then be no more unexpected than that between any other upward and downward selected lines in a two-way selection experiment.

But this mechanism could only operate in so far as the previous selection had led to important progress. The lack of divergence produced in our Bar experiments is probably due to low response to selection in the early generations, and to the fact that such response as did occur seems to have been asymmetrically in one direction, towards large eyes.

Thoday also claims that in his experiments a considerable degree of mating preference was developed, so that both high and low flies tended to mate predominantly with their like. Clearly any such tendency would increase the rapidity of divergence. It is surprising, however, that the experiments involved sufficient selection pressure in this direction to bring about the very striking effects Thoday claims. Other attempts to select directly for mating preferences [7, 8, 9] have made much slower progress. It is most desirable that Thoday's results, both on the achievement of rapid divergence, and particularly on the development of mating-preferences, should be repeated, with special attention to the strict virginity of the flies used, which is obviously a matter of crucial importance. In experiments on a similar plan, with selection for high or low expression of cubitus-interruptus-Dominant, Scharloo [10] found that disruptive selection led to a considerable increase in phenotypic variance, but no divergence into two phenotypic groups, and he found no evidence of the development of mating-preferences.

The most interesting fact about disruptive selection which has emerged from the experiments reported here is the greater increase in phenotypic variance in the anti-canalization as compared with the canalization lines. In the canalization lines, selection has been exerted against the effects of the two different environments. It would therefore be expected that the partition of the phenotypic variance into genotypic and environmental components would be shifted in favour of the former, while in the anti-canalization line an opposite result would be anticipated. The phenotypic variances in constant (or nearly constant) environments which were estimated at the end of the experiment will be expressions of the genetic variation for large and small eye-size available in the population, the buffering of individual development against minor environmental fluctuations, and the buffering against genetic variance, which may be expected to be closely comparable to that against environmental factors. Since in the canalization line, the environments were acting to suppress the differences in eye size which were selected, while in the anti-canalization line they were acting to produce them, it might, on the simplest assumptions, be expected that the canalization line would finish up with more genetic variation for eye size than the other. However, it would also be expected to be—and in fact has been found to be—better buffered against environmental variations, and therefore probably against genetic variation also. The actual result which has been described above seems to show that the latter effects have in practice proved more important than the former in determining phenotypic variance. It seems doubtful whether there is any theoretical reason why this should always, or even usually, be the case. A similar experiment with a different foundation stock might have led to the opposite result. In fact a comparison of canalization line no. 4 with anti-canalization line no. 9 in table 2 provides an example of behaviour exactly the opposite of that shown by the means of all the lines.

There is, therefore, no major generalization to be drawn from the results of the disruptive selections; but these experiments are, perhaps, of significance in demonstrating that the theory of the process is considerably more subtle and complex than current treatments admit.

SUMMARY

Starting from a genetically variable stock homo- or hemi-zygous for Bar two selection lines were set up, one selected for decreased sensitivity to the effect of larval temperature on eye size ('canalization line'), the other for increased sensitivity ('anti-canalization line'). In each generation a sample of larvae was grown at 25°C throughout life and another sample at 25°C for the first 48 hours, followed by 18°C until emergence. In the canalization lines a selection was made of individuals (five males and five females out of 100) least affected by the temperature treatment and in the anti-canalization lines for a similar number most affected by the treatment. These ten males and females were allowed to mate at random and from the eggs produced random samples were then treated in the next generation in a similar manner in the two temperatures. Precise counts of facet numbers on the right eyes were made at generations 4 and 9 and it was clear that selection had been effective both in decreasing and increasing temperature sensitivity.

The whole canalization line can be regarded as a population which has been subjected to disruptive selection for two different criteria: one (small eyes) being regarded as adaptive to the low-temperature regime or habitat; the other (large eyes) being regarded as adaptive to the high-temperature habitat. A similar type of analysis can be applied to the anti-canalization line in which, however, the selective values of the phenotypes are regarded as reversed in value in the two habitats. The experimental procedure employed involved not only random mating between individuals selected according to these two criteria but also random allocation of the offspring of this panmictic population to the two habitats of the next generation. This corresponds to a natural population in which there is disruptive selection exerted by two different habitats but no habitat preferences exhibited by members of the population. Under these circumstances, in our experiments, disruptive selection produced only rather slight increases in phenotypic variance, which were rather larger in the anti-canalization lines. Possible reasons for the difference between this result and those reported by Thoday and Gibson are discussed, and it is suggested that an important reason may be the slowness of the response to selection in the first few generations in our lines.

Acknowledgement
We would like to thank Dr B. Woolf for advice and assistance in the statistical treatment of the data.

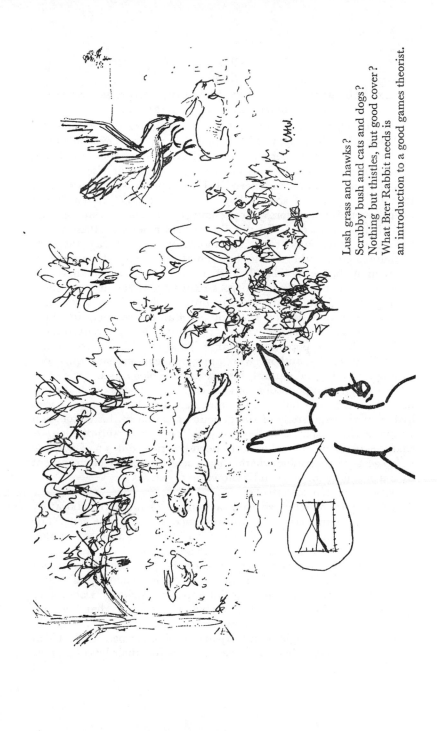

Lush grass and hawks?
Scrubby bush and cats and dogs?
Nothing but thistles, but good cover?
What Brer Rabbit needs is
an introduction to a good games theorist.

14

ENVIRONMENT SELECTION BY DROSOPHILA MUTANTS
(WITH B.WOOLF AND M.M.PERRY)

In the development of the mathematical theory of evolution by Haldane, Sewall Wright, Fisher and others, the environment is normally treated as uniform and the behaviour of various mutants in respect to it is discussed simply in terms of a coefficient of selection by which each mutant is characterized. In practice it is, of course, clear that the majority of species live under conditions in which the environment is not uniform but presents many minor variations. The variant types present in the population will react with these variations of the environment in much more complicated ways than are contemplated by the simple mathematical theory. One type of interaction between animal and environment which may be expected is that different genotypes may have different preferences among the variants of environment open to them. There may then be a selection by a particular genotype of an environment in which it will be subjected to selective pressures somewhat different to those which apply to the population at large. This possibility has frequently been discussed by authors who approach evolutionary questions through a study of natural history and ecology (see review by Thorpe[1]), who point out that different sub-species or local races may show different habitat preferences. There has, however, been little experimental study of the genetic variation in this connection within a single species not already systematically sub-divided. A small experiment has therefore been carried out on *Drosophila melanogaster* to determine whether such selection of the environment by particular genotypes can be shown to occur.

MATERIALS AND METHODS

The genotypes investigated were a number of standard laboratory stocks each characterized by the presence of one major gene. The general genetical background of the various stocks was presumably considerably diversified and the behaviour of the stock is therefore not necessarily a direct consequence of the presence of the major gene after which the stock is named. The environmental preferences of the stocks were tested in a special apparatus made of wood. This was octagonal in shape and consisted of a central chamber, approximately 11 in. in diameter, which communicated by holes, 1 in in diameter, with eight regularly-arranged chambers each 16 in long and 4 in across their narrower, 18 in across their broader ends.

The height of the whole apparatus was 12 in. In these eight chambers a series of environments were set up by varying the factors of temperature, light and humidity. The control of these variables was somewhat crude. During the experiments the apparatus was kept in a darkened room and light was provided by a 1000-watt bulb hanging 4 feet above the central compartment. The sectors of the apparatus which were to remain light were provided with glass lids and painted with white paint. The dark ones had wooden lids and were painted black. The dry and damp environments were obtained by putting a mass of soaked cotton wool in the damp compartments and an open Petrie dish containing calcium chloride in the dry ones. The warm compartments were electrically heated. The heating system had a thermostatic control, but owing to the low degree of thermal insulation provided by the equipment the temperature did not remain constant to closer limits than two degrees Centigrade. Each compartment contained a dish of an alcohol-acetic acid mixture to attract the flies.

The flies to be used in the experiment were counted out the day before. At the start of the experiment a number of different populations were shaken into the central compartment without etherisation. They were then left for a given length of time during which they were able to pass through the holes communicating from the central compartment with the various environmental sectors. After the lapse of the appropriate time the flies in the various sectors were etherized and counted.

RESULTS

Some results obtained with this apparatus have been mentioned previously [2]. In these first experiments the apparatus was in a room open to daylight and the control of the other environmental variables was exceedingly poor. There was therefore a considerable lack of homogeneity between successive experiments using the same stocks of flies. The results of individual experiments suggested that there were differences between various stocks in their response to light and dark, but the heterogeneity between replicate experiments made it impossible to draw any firm conclusions.

The results of more recent experiments using the somewhat improved conditions mentioned above are shown in table 1, which gives the results of three successive experiments using Wild Type, rough, aristaless, purple, apricot, black, and forked. Time allowed before counting in the respective experiments was 6 hours, 5½ hours, and 5 hours. The numbers moving consequently vary from experiment to experiment. No attempt has been made to analyse the data on mobility. Attention has been confined to the destinations of the flies that moved.

Table 1. Results of triplicate experiments with seven stocks.

Stock	Expt.	Total flies	light dry cold	light dry hot	light wet cold	light wet hot	dark dry cold	dark dry hot	dark wet cold	dark wet hot
						Destinations of moving flies				
+	1	160	7	8	11	10	2	1	4	2
	2	200	20	13	13	17	18	3	3	4
	3	180	18	13	20	19	7	3	4	8
	Total	540	45	34	44	46	27	7	11	14
ro	1	170	20	7	18	4	–	1	3	1
	2	200	35	4	12	2	3	–	2	1
	3	200	35	9	28	12	14	3	15	1
	Total	570	90	20	58	18	17	4	20	3
al	1	110	2	3	9	9	1	–	1	–
	2	150	9	3	8	16	2	1	6	–
	3	200	10	8	18	20	2	1	–	1
	Total	460	21	14	35	45	5	2	7	1
pr	1	120	14	11	13	8	–	–	1	–
	2	200	9	8	13	13	3	1	–	–
	3	200	13	12	20	14	2	1	–	–
	Total	520	36	31	46	35	5	2	1	0
wa	1	100	7	7	7	4	–	–	1	–
	2	200	17	9	9	5	1	–	1	–
	3	200	19	21	21	11	2	1	3	3
	Total	500	43	37	37	20	3	2	5	3
b	1	200	12	13	23	12	1	2	3	1
	2	200	17	16	24	14	2	–	1	–
	3	200	15	9	40	16	6	1	4	2
	Total	600	44	38	87	42	9	3	8	3
f	1	200	10	6	17	12	3	–	8	–
	2	200	12	5	17	10	3	3	3	–
	3	200	18	5	21	16	5	4	5	10
	Total	600	40	16	55	38	11	7	16	10

The first question is whether the differences in the distribution of the flies of any one stock in the three experiments are of an order to be expected by chance deviations from a consistent average, or whether they indicate real differences in conditions between the experiments. This can be tested by the χ^2 method. For each stock there are eight choices and three replicates, allowing the calculation of a consistency χ^2 with 14 degrees of freedom, except for the *pr* stock, where the absence of any flies in one of the chambers cuts the degrees of freedom to 12. All the stocks but one—the *ro* stock— gave a consistency χ^2 below the conventional ·05 probability significance point, and the summed χ^2 for 96 degrees of freedom was 113·5, corresponding to a probability of about one in ten. Thus the triplicate experiments may be regarded as reasonably self-consistent. It is therefore legitimate to take the totals for the various stocks, as shown in the table, and to apply the χ^2 method to them to test for differences in behaviour between the stocks. There are 42 degrees of freedom available, and the χ^2 value is 181·9, which corresponds to a probability much smaller than 10^{-20}. It may therefore be taken as proved that there are real differences in choice of destination between the flies in the various stocks.

That proves the main point, but it is of interest to look into subsidiary questions. Characterization of behaviour differences would be much simplified if the proportion of flies choosing hot rather than cold were the same for wet and dry or for light and dark environments, and if, similarly, the choice of relative humidity or luminosity were not influenced by the other two variables. In such circumstances, the behaviour of each stock would be fully specified by three parameters, the percentages preferring respectively light to dark, dry to wet, and hot to cold. Table 2 gives the deviations between the observed totals for each stock, as shown in table 1, and the number expected on the assumption that choice criteria are independent. Thus 45 wild type flies went into the first chamber, against an expected number of 46·7, giving a deviation of $-1\cdot7$; and similarly for the other figures. Since for each stock there are eight observations, and the hypothesis requires four parameters—the total of flies moving, and the three preference ratios—conformity to the hypothesis for each stock can be tested by χ^2 for four degrees of freedom. Only one of the stocks, the wild type, gave a value significant at the ·05 level, but the total χ^2, for 28 degrees of freedom, was 46·48, corresponding to a probability of less than ·02. This is entirely due to the larger χ^2 value, 14·08, for the wild type. The most just conclusion seems to be that some interaction between choice

criteria is probable, though not convincingly demonstrated; but that
to a first approximation the three criteria of choice can be considered
as more or less independent.

Table 2. Deviations from expected numbers on the hypothesis of
independent choice criteria.

Stock	light dry cold	light dry hot	light wet cold	light wet hot	dark dry cold	dark dry hot	dark wet cold	dark wet hot
+	−1·7	−3·1	−3·5	+ 8·2	+10·7	−6·0	−5·6	+0·8
ro	+4·8	−0·7	−6·4	+ 2·3	− 3·2	−0·9	+4·8	−0·7
al	+1·6	−3·7	−5·7	+ 7·9	+ 2·5	−0·3	+1·7	−3·8
pr	−3·6	+0·4	+2·1	+ 1·1	+ 2·9	+0·3	−1·4	−1·8
wa	−2·6	+5·4	+1·7	− 4·5	− 1·0	−1·8	+1·9	+0·9
b	−9·6	+6·9	+7·2	− 4·4	+ 3·2	−0·4	−0·7	−2·1
f	+3·9	−5·0	−3·1	+ 4·2	+ 0·3	+0·8	−1·2	0·0
Total	−7·2	+0·1	−7·7	+14·9	+15·4	−8·2	−0·4	−6·7

Table 3 gives the analysis of choice criteria on this assumption of
independence. It will be seen that there is no significant difference
between stocks with respect to temperature preference, but the
behaviour with respect to humidity and still more to luminosity
does show very significant variation.

Table 3. Relative numbers moving according to each criterion of choice.

	+	ro	al	pr	wa	b	f	Total	χ^2 6d.f.	P
Total moving	228	230	130	156	149	234	193	1,320		
Light	169	186	115	148	137	211	149	1,115⎫	31·392	≪·001
Dark	59	44	15	8	12	23	44	205⎭		
Dry	113	131	42	74	84	94	74	612⎫	20·483	<·01
Wet	115	99	88	82	65	140	119	708⎭		
Cold	127	185	68	88	88	148	122	826⎫	9·669	>·10
Hot	101	45	62	68	61	86	71	494⎭		

The results of another series of experiments with wild type and six
other mutant stocks are shown in table 4. In the first two experiments, denoted L1 and L2, the central chamber was lighted, and in
the other two, designated D1 and D2, it was kept dark. It is obvious
at a glance that in the latter pair of trials there was a much smaller
migration to the dark destination. One stock, *app*, was not represented in L1, and another, *v*, was absent in D2.

Analysis was done as before, with L1 and L2 considered as one set of duplicates and D1 and D2 as another set. In the D series, the numbers migrating to the four dark chambers were so small as to give expectations for which χ^2 values would be suspect, so analysis was confined to the flies migrating to the four lighted chambers. In the D series, the consistency χ^2 value, for 18 degrees of freedom, was 16.58, corresponding to a probability of just above 0·5, indicating that conditions were under adequate control. For heterogeneity between stocks, the χ^2 was 34·76 for 18 degrees of freedom, significant beyond the ·01 point. The D series therefore confirm the conclusion that mutant stocks differ in their choice of environment. In the L series, the consistency χ^2 was much larger than would be expected

Table 4. Results of two sets of duplicate experiments with seven stocks.

| | | | Destinations of moving flies | | | | | | | |
Stock	Expt.	Total flies	light dry cold	light dry hot	light wet cold	light wet hot	dark dry cold	dark dry hot	dark wet cold	dark wet hot
+	L1	200	16	3	15	18	7	2	2	2
	L2	210	10	10	17	25	–	8	5	15
	D1	160	7	9	18	18	–	–	–	3
	D2	200	11	11	30	22	1	3	1	5
ca	L1	200	16	17	22	27	3	–	2	–
	L2	210	21	17	29	26	2	2	–	3
	D1	170	12	24	20	26	1	–	–	2
	D2	180	9	8	22	20	2	1	–	–
w	L1	180	19	8	7	13	7	3	6	5
	L2	200	18	11	7	10	4	4	8	9
	D1	200	18	20	15	26	–	1	1	9
	D2	200	5	13	15	11	6	2	8	10
e	L1	200	16	12	18	23	–	–	2	1
	L2	200	22	26	15	37	–	2	2	6
	D1	200	22	23	24	36	–	1	–	1
	D2	140	18	19	21	18	1	3	3	7
v	L1	200	16	9	22	34	11	–	8	–
	L2	150	19	8	18	5	8	2	12	9
	D1	200	24	13	34	24	1	1	1	–
stw²	L1	140	5	4	4	5	3	–	–	–
	L2	170	11	7	13	15	–	–	4	1
	D1	200	3	7	13	16	–	1	–	1
	D2	200	7	5	14	28	1	2	–	1
app	L2	210	21	13	17	18	4	2	4	3
	D1	120	7	8	9	11	–	1	3	–
	D2	200	7	5	8	9	1	–	4	–

by chance, being 86·59 for 40 degrees of freedom, significant beyond the ·00001 point. The discrepancies were due to the wild type and v flies, all the others giving low χ^2 figures. Between stocks the χ^2 was even greater, being 135·76 for 42 degrees of freedom. Even allowing for heterogeneity between duplicates, this is almost significant at the ·05 level, as is shown in table 5. As far as it goes, the L series therefore also tends to support the conclusion already reached. But since the main effects are not irrefutably established in the second set of experiments, it is not worth while analysing the data for secondary features.

Table 5. Test of significance for L series.

	Degrees of freedom	χ^2	Mean square
Between stocks	42	135·762	3·232
Between experiments	40	86·687	2·167
Ratio			1·644
·05 Point, F_{40}^{42}			1·665

DISCUSSION

Although the two sets of experiments described above were of a preliminary nature, they have given definite evidence that different stocks of *Drosophila* exhibit definite and particular preferences when offered a choice of environment. Any attempt to study the physiological basis of the choices and the way in which they are determined by the effects of the mutant genes concerned would require much more complicated equipment and a more precise control of the conditions than were available. It is not intended to pursue such investigations in the immediate future. The experiments were intended merely to provide evidence of whether such environmental preferences actually exist or not.

Jentzer and Ludwig[3] have recently described an experiment in which various mutant races of *Drosophila* were released in a rectangular box 1 metre long, one end of which was transparent and illuminated by a 40-watt lamp, while the other end was covered with dark paper. They investigated the position taken up by the flies five minutes after the beginning of the experiment. They found differences in this respect between the different stocks used, and in general came to the conclusion that those with dark body colours or dark eye colours tended to seek the light more than the others. This conclusion does not hold in our experiments, in which both dark eye colours (purple), dark body colours (black) and pale colours (apricot) all tended to shun the dark, as did aristaless, in which neither body nor eye colour are abnormal. The differences between our results and those of Jentzer and Ludwig are probably to be explained by the different experimental conditions. It is noteworthy that in their experiments the flies tended to be concentrated at the two ends of the box. This may indicate, as it appears to do at first sight, a tendency to go to the extremes of light or dark available; but perhaps the behaviour of the flies is influenced by the fact that at the ends of the box there are more solid surfaces within a given amount of space than there are in the middle. In our apparatus the structural complexity was similar in all different environments. Moreover, the difference in intensity between the light and dark holes leading out of the central compartment was probably greater than the contrast available in Jentzer and Ludwig's apparatus.

Perhaps the best previous evidence for the existence of habitat preferences among mutants of the same species is that described by Hovanitz[4, 5] in the butterfly *Colias eurytheme*. The females of this species exist in two forms, an orange and a white. The white phase is more active during the early part of the day, while later on the activity of the orange phase increases. The two forms are interbreeding, and co-exist in the same populations, the proportions differing according to the elevation and latitude of the territory studied. It is not clear whether the physiological difference in the activity of the two forms leads to any assortative mating or not.

It will presumably often be the case that when two forms within a population exhibit habitat preferences the selective values of the different genotypes will depend on the environment in which they spend their lives. A full discussion of the possible evolutionary consequences of this would involve much intricate mathematical theory, which the factual material available is hardly yet rich enough to justify. Certain points, however, emerge from quite simple considerations.

There are two opposite extreme cases. In the one, we may suppose that the whole population remains panmictic, and that there is free migration to the environment of choice open to all individuals. Then the fitness coefficients of the various genotypes will be composite functions of the preferences for, and the selective values in, the different environments; and so long as these remain constant, natural selection will follow the course usual in a population with constant fitness coefficients. If, at the other extreme, mating is strictly within the group which has chosen a particular environment, and there is no migration between environments, we shall have two isolated sub-populations which will evolve each according to its appropriate fitness coefficients; if each genotype is favoured in the environment it tends to choose, selection will eventually proceed to the fixation of different genes in each sub-population. A more interesting situation arises in the intermediate case, when there is a certain amount of migration between two sub-populations in which selection is proceeding in opposite directions. Under some circumstances of this kind, the gene frequencies in one of the sub-populations might reach an equilibrium between the opposing forces of selection and migration. Moreover the equilibrium within a group will be stable provided that a chance decrease in the frequency of the favoured gene leads either to a decrease in migration or to an increase in the effectiveness of selection—a situation which is biologically not implausible.

The occurrence of habitat preferences, with different fitness coefficients in the different environments, and restricted intermingling between the sub-populations, may therefore lead to the appearance of a stable polymorphism in the population as a whole. The stability will in practice probably not be absolute, since in such a situation one would expect both the habitat preferences and the fitness coefficients to undergo gradual change; but it remains true that two forms exhibiting different habitat preferences could co-exist for a much longer time than would be possible if their ecological behaviour were the same.

In the experiments which have been recorded above there is no evidence that the various races have chosen those environmental situations in which they would be most favoured by natural selection. It may well be the case that in nature different mutants do not always choose the optimum conditions for their existence. It is, of course, only when the choice is such that it improves the chance of survival of a mutant which would otherwise be eliminated that the selection of the environment by the mutant plays an important part in evolution. It must also be remembered that the choice of habitat is made, not directly by the genotype, but by the phenotype, which has itself been influenced by environmental conditions during development. We may, therefore, be confronted by a subtle system of mutual interactions between the genotype, the effects of the environment on development, the natural selective forces and the habitat preferences of the phenotype, which can bring into play the processes of 'organic selection' [6] or 'genetic assimilation' [7].

SUMMARY

1. An apparatus was constructed in which a central compartment communicated with eight wedge-shaped chambers. The conditions in these chambers were rather roughly controlled so that each of them was either hot or dry, cold or wet, light or dark. Mixed groups of adult *Drosophila melanogaster* were released in a central compartment and their distribution among the various chambers investigated some hours later.

2. In a first group of three experiments the stocks used were Wild Type and six others, each homozygous for a well-marked recessive gene; the genetic backgrounds were not standardized. When the numbers which had moved into the various compartments were analysed it was found that there was little evidence of heterogeneity between the various experiments, but overwhelming evidence of differences in behaviour between the stocks. In another series, using Wild Type and another six mutant stocks, there was also strong heterogeneity between stocks but in this case some significant heterogeneity between experiments. Although this second series is therefore not as satisfactory as the first, it provides confirmatory evidence for the existence of differences in behaviour between different mutant stocks.

3. No attempt has been made to analyse in detail the physiological reasons for the different choices made. The facts agree to a first approximation with the hypothesis that the criteria of choice with respect to heat, light and humidity, are independent. If this hypothesis is adopted, the figures indicate that there is no significant difference between stocks with respect to temperature preferences, but that behaviour with respect to luminosity and humidity shows significant variation.

4. It is pointed out that in a heterogeneous environment preferences of particular mutants for those environmental conditions in which they were most favoured by natural selection might under some circumstances lead to the maintenance of a relatively stable polymorphism. [1954]

15

Two mechanisms have been advanced for the origin of reproductive isolation between species. Muller[1], dealing in the main with barriers to crossing in the later stages of species divergence, such as hybrid inviability and infertility, suggests that these arise almost by chance as a product of change in the genetic background either by genetic drift or as adaptation to different biological situations. This would lead to accelerating divergences as the process continues, or, as Muller puts it, 'ever more pronounced immiscibility as an inevitable consequence of non-mixing'. Dobzhansky's suggestion[2], which is perhaps complementary rather than antagonistic to Muller's, is that when sufficient divergence between two species has arisen so that the hybrids are less well adapted for any available habitat then either parental type, there will be selection for sexual isolation. That is to say — if mating can take place and if the resulting hybrids are inviable or infertile, then natural selection will operate to reduce the chance that mating will occur, either by reducing the chance of encounter or the chance of mating with members of the other species when they are encountered.

Some writers, in discussing the mechanism proposed by Dobzhansky have suggested that 'natural selection will favour any mechanism which prevents the wastage of gametes involved in unsuccessful hybridization'. This seems to be unduly teleological. Natural selection will only tend to suppress crossbreeding if those individuals which hybridize will in consequence pass on fewer gametes in the form of pure-bred offspring. It would seem probable that this would be more often the case in females than in males. In *Drosophila melanogaster*, for instance, females seem reluctant to mate again for a period of two or three days after an effective mating. If the first mating has been heterogamic, this will reduce the number of purebred offspring that she will produce in her lifetime. Gestation in mammals will have a similar effect. But the male, who must on the average have the same number of effective matings in his life as the female, is usually capable of many more if willing females are available. It follows then that willingness to cross-breed, which may merely be a sign of greater general sexual activity, will not necessarily reduce the number of purebred progeny that a male will leave. If Dobzhansky's mechanism for the establishment of sexual isolation is correct, it follows that it should be in the main a matter of female preference. Merrell[3] has recently presented evidence that it is the female which exercises discrimination in matings between *D. pseudoobscura* and *D. persimilis.*

Koopman[4] has shown that selection leads to an intensification of the sexual isolation between these two species. Using marked stocks of the two species, he selected continually for pure-bred flies — the progeny of parents that had mated homogamically. He showed that the proportion of hybrids emerging declined dramatically after a few generations of selection. More recently, Wallace[5] and King (private communication) attempted to demonstrate the production of sexual isolation by selection within a species. They used two stocks of *D. melanogaster,* from widely separated localities, which had each been marked by a different recessive gene. After twelve generations, when the experiment was first reported, little change in the proportion of wild-type flies emerging had been observed, but in subsequent generations the proportion declined significantly, showing that sexual isolation had been to some extent established. This was confirmed by observation of individual matings.

Our own experiment on very similar lines was started before we were aware of Wallace's work, and as our work was slightly different in conception, we decided to proceed with it. In Wallace's experiment, the mutants were used solely as markers, the stocks because of their origin presumably differing in many genes. As it happens, we had used in our work stocks marked with the autosomal recessive genes, ebony and vestigial, which has been extracted from a population in which the two had been segregating for many generations. The original stocks making up this population were actually those used by Rendel[6] in his work on the effect of light on the mating of these mutants. Our two foundation stocks, both of which contained a considerable amount of genetic variability, were thus probably very similar except for the marker genes. These genes were chosen because of the ease of scoring but they do react differently to light and, as Rendel has shown, ebony males mate more frequently in the dark than in the light.

In the first experiment of this type that we carried out, there appeared in the seventh generation some flies that were both ebony and vestigial, indicating that in previous generations either a non-virgin female or else a wild-type heterozygote had been used as a parent with the result that each mutant stock was contaminated with the other gene. Theoretical consideration of the effect of this showed that the proportion of double recessive flies should increase by a factor of four each generation until they reached a level of 11% of all flies emerging. At that point, the proportion of flies in each mutant stock that were heterozygous for the other gene would be two-thirds. There would then be a continual interchange of genes between the two stocks. In addition, one-third of the apparently pure mutants used as parents would be derived from heterogamic matings, thereby reducing the selection for sexual isolation. We therefore discarded the line and started afresh with stringent precautions against non-virginity, parents being collected over a seven hour period. In the two experiments presented here in detail, no double recessive flies were ever observed.

DESCRIPTION OF EXPERIMENTS

Box experiment

Two mutant strains of *D. melanogaster* homozygous for the genes ebony and vestigial respectively were used. They had been extracted from the same population, after segregation for many generations. At the start, 54 males and 54 virgin females from each of the stocks were put together into a breeding box (size 18 in × 18 in × 7 in) which contained ten unstoppered quarter-pint bottles of maize meal-molasses-yeast-agar medium. Flies were etherized for counting, but were not put into the box until three hours after complete recovery from anaesthesia. The box was then placed in a constant temperature room at 25°C. All phases of the experiment were done at this temperature. The box was always put in the same part of the room, where, due to the direction of the light, two sides of the box near the edge were in slight shadow. The ebony flies, immediately the box was positioned, migrated towards the light source, that is, towards the shaded edge. After some time, the majority of them moved more freely about the cage.

After six days of mating, the ten food bottles were removed, cleared of any flies which remained inside, and stoppered. The parents were discarded. The count of the next generation was started five days afterwards, i.e. on the eleventh day after the parents were put into the box. Three types of flies emerged; hybrids from heterogamic matings, and the two mutants ebony and vestigial from homogamic fertilizations. For $3\frac{1}{2}$ days every fly which emerged was counted. The culture bottles were completely cleared at 10 a.m. Flies which emerged by 5 p.m. on the same day were segregated and mutants were kept in separate vials to be used as parents for the next generation when one to four days old. When insufficient virgins were obtained, those collected were bred with their own kind, and the experiment carried on from their progeny. In the box experiment this was done three times in 38 generations.

In order to ensure that any changes in external conditions had not affected the course of the experiment, controls were done on the box experiment in the later generations. Parent virgin flies were obtained from the original stocks and put into a box of identical proportions to the experimental one. The control box and the experimental one received exactly the same treatment throughout. This was done seven times between the 25th and 35th generations.

Jar experiment

An experiment on similar lines was run in conjunction with the cage one. A two-pound glass jar containing approximately one inch of food was used as the breeding chamber. The number of parent flies employed in this case was between 20 and 30 of each sex of the mutants. Again, it was sometimes necessary to mate the virgins with their own kind to produce sufficient numbers for the next preferential mating. This was done three times in 33 generations. From generations 1 to 12 the parent flies were still under ether when put into the jar, as it was thought that they might otherwise escape. This was found, however, to be unsatisfactory. So from the 13th generation onwards the parents were introduced into the jar three hours after recovery from the ether. The jar was put into the same constant temperature room and at the same time as the box. Thereafter, all operations, such as clearing parents from the jar, counting and segregating flies of each generation, etc., were carried out at the same time and in an exactly similar manner to the box experiment.

Because of the small capacity of the jar compared with that of the box and the fact that there was little or no variation in the light within the jar, it was assumed that any tendency towards an ecological isolation between the ebony and vestigial flies would be eliminated.

RESULTS

One of the first impressions at the start of the experiments was of the great fluctuation in results from generation to generation. The jar experiment was in fact started to try to remove this by having all flies developing in one food mass. Our criterion of isolation has been the ratio of wild-type flies, produced by heterogamic matings, to the total number of mutants produced by homogamic matings. The standard deviation of this ratio due to chance fluctuation, determined from the mean square difference between successive generations, was 0·16 for the box experiment and 0·24 for the jar. This fluctuation is equal to that produced by random sampling of 160 units and 70 units respectively from a population made up of two types of objects with equal frequency. The total count was actually of the order of 2000 flies in both cases. But the number of female parents was 108 and 50 in the box and jar respectively. The observed fluctuations suggest that the effective units are the initial inseminations of the individual females. In this respect, it is of interest that of the individual platings of females taken from the box after six days of mating, 660 gave offspring all of the same type and only 75 had mixed offspring. However, whatever the reason, it is still true that too many flies were counted each generation and a sample of a quarter of the size that we took would have been quite adequate.

Box experiment

The results of the box experiment are set out graphically in figures 1 and 2. The graphs are moving averages over five generations to smooth out fluctuations. In figure 1, the number of hybrids is expressed as a percentage of the sum of the ebony and vestigial emergences. Figure 2 shows separately the numbers of the three types of flies emerging. From the first to the eighteenth generation a more or less steady decline in the percentage of hybrids is noted. The lowest percentage of hybrids in any individual generation was 10·3% at the eighteenth, with emergences of $++246$, $e\,736$, $vg\,1640$. Only once afterwards, at the 23rd generation, does the vestigial line graph fall as low as the control mean for this mutant. Thereafter the values remain high for vestigial emergences. The hybrid figure drops, and is lowest between the 16th and 18th

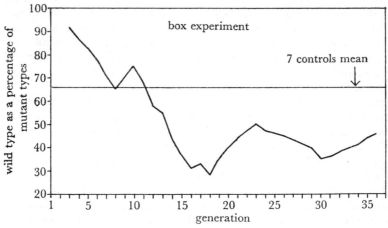

Figure 1. Results of box experiment. Number of hybrids expressed as a percentage of ebony and vestigial emergences.

generations, only rising a little and slowly towards the end of the experiment. During the whole 38 generations the emergence values for ebony alternate slightly above and below the figure for the control mean. This suggests that the sexual isolation, after the 18th generation, is due mainly to the increase in the number of homogamic matings of the vestigial flies.

The average values for the seven control generations are also shown in figures 1 and 2. The proportion of wild-type flies to mutants averages 0·66, compared to the proportion in the selected population at the same period of 0·38. The figures for the individual mutants show that the change is due to a decrease of wild-type flies and an increase of vestigial.

It has been shown by Rendel[6] that ebony reacts to light intensity in its mating behaviour. It seemed possible that the sexual isolation was due to an accentuation of this response. Towards the end of the experiment, therefore, duplicates of the selection box were made up from parents from the selected stock but were kept instead in complete darkness. The ratio of wildtype to mutant offspring was 0·48 compared to 0·46 for the three contemporary generations in the light. It seems therefore that the demonstrated sexual isolation is not concerned with phototropic response. However,

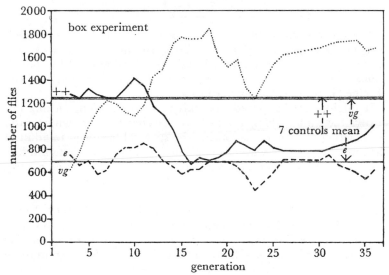

Figure 2. Results of box experiment. Numbers of three types of flies emerging shown separately.

there were many more ebony flies in the dark boxes—in fact the average of the three tests (1102 flies) had only once been exceeded by a single generation in the light, and the average in the last few generations of the latter was about 650. There was correspondingly a shortage of *vg* flies, but the proportional effect was not so great. This agrees with Rendel's observation that ebony males show greater sexual activity in the dark.

Between the 20th and 30th generations, the females were placed in individual vials after they had been removed from the box, and their progeny were examined on emergence. This was done with six generations of the selected stocks and with three of the controls. The results in terms of effective matings are given in table 1.

Table 1.

Control females	Inseminated by	
	e	*vg*
e	71	69
vg	41	63
Selected stocks		
e	151	108
vg	77	142

There is a slight tendency to homogamic mating in the controls but the heterogeneity χ^2 is only 2·15. In the selected stocks the tendency is much more marked and the χ^2 value is 27·14. This is confirmatory evidence that some degree of sexual isolation has been obtained in the selected population.

It might have happened that this type of selection, picking out always the mutant flies, would have affected the segregation ratio by selecting those genes favouring the survival from egg to adult of the mutant types. However, a check based on several thousand flies at the end of the experiment showed no differences between control and selected stock in the segregation ratio for either mutant.

Jar experiment

The results for the jar experiment are given in figures 3 and 4. In the jar, the light intensity was much more uniform than in the box and in addition the flies were more confined—the volume of the jar being in the order of one-fiftieth of that of the box. Here again there is a decline in the proportion of wild-type flies as the experiment proceeds, although the proportion at the end is a trifle higher than in the box experiment. However, the ratio of wild-type to mutant

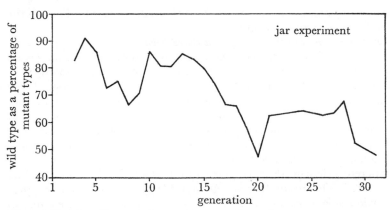

Figure 3. Results of jar experiment. Compare figure 1.

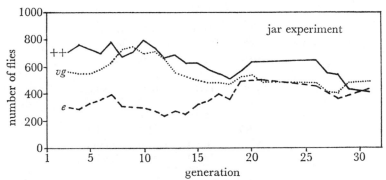

Figure 4. Results of jar experiment. Compare figure 2.

flies has declined from 80% to between 50% and 60%. As was noted above, there was a change in method in the middle of this experiment. Up to generation 12, flies were put into the jar etherized but afterwards they were put in in an active state. This change does not appear to have affected the sexual isolation. In the five generations before the change, the average ratio of wild-type to mutants was 0·86 and in the five after the change it averaged 0·80. It may however have affected the separate types. The numbers of *vg* and wild-type flies decline by about one-third as a result of the change, whereas the ebony count is unchanged. The subsequent change in the wild-type/mutant ratio appears in this case to be due to an increase in the number of ebony flies. Control experiments were not carried out on the jar population, as the latter was subsidiary to the main experiment in the box.

Sexual preferences in inbred lines and closed populations

It is convenient to present here a small amount of data on sexual preferences between lines and populations chosen at random, with a bearing on the 'chance' occurrence of sexual isolation. These experiments were carried out by the usual 'male choice' method in which males are given equal numbers of two types of female, one of which is recognizable—in our case by a spot of silver paint. The females are then examined for the presence of sperm in the seminal tract. In the first case, two wild-type inbred lines of completely different origin were used, given in our stock list the symbols W20 and K7. The results are given in table 2, which shows the proportion of females inseminated. In all cases, the males were equal in number to the females of each of the separate lines.

Table 2.

females marked	males	W20 females	K7 females	Duration of mating
W20	W20	22/23	11/25	40 mins.
	K7	17/23	9/25	40 mins.
W20	W20	17/19	16/20	70 mins.
	K7	4/20	1/19	70 mins.
K7	W20	9/10	5/10	30 mins.
	K7	8/14	9/14	60 mins.
K7	W20	10/20	2/19	30 mins.
	K7	10/20	5/20	45 mins.

A similar experiment was then done using lines which, although of common origin, had been selected in different directions without inbreeding for 20 generations for number of chaetae on the fourth and fifth abdominal sternites. There was no overlap in chaetae count between the high and low lines used, so that this character could be used for identifications. The results are given in table 3, for matings between one high line, H1 and two low lines, L4 and L5.

In these rather meagre results, there is little suggestion of sexual isolation having developed by chance in either the inbred experiment or in that with the selected lines. In the latter, it is of interest that the selection for the quantitative character has caused a differentiation in mating ability. The H1 males are poorer than those from the two low lines but on the other hand the high females seem to be better. But this seems to be a general change in sexual drive, not specifically adapted to the other sex of the same line. Wallace[7](and personal communication) has also tested whether the mere isolation of two populations is sufficient to cause sexual isolation to arise between them. His populations had been separated for 80–100 generations, and differed in certain morphological characters (primarily abdominal pigmentation). In an extensive series of tests, no tendency towards preferential mating could be detected nor does there seem to have been any evidence of differences in the intensity of general sexual drive.

Table 3.

males	H1 females	L4 females	Duration of mating
H1	3/10	1/9	30 mins.
L4	9/14	3/14	30 mins.
	H1 females	L5 females	
H1	11/20	5/19	60 mins.
L5	20/24	17/25	60 mins.

DISCUSSION

Our results may be summarized in the statement that some sexual isolation developed when we selected for a tendency towards homogamic matings, but that none was found to have arisen by chance in a few lines which had been selected for abdominal chaeta number or inbred. Laboratory experiments on evolutionary mechanisms can, of course, only be indicative and not demonstrative—they can show what might happen in wild populations, rather than what has happened. As far as they go, our experiments lend support to the mechanism suggested by Dobzhansky rather than that discussed by Muller. But when attempting to apply these results to occurrences in nature, one must bear in mind the ways in which artificial populations may fail to imitate conditions in the wild.

It is perhaps misleading to put Muller's hypothesis of the chance origin of sexual isolation in antithesis to that of Dobzhansky, which attributes it to the action of selection. In all probability, both mechanisms have operated in the wild in different cases. Some evidence supporting Dobzhansky's hypothesis comes from Dobzhansky and Koller[8] who found, in an analysis of crosses between *Drosophila pseudoobscura* and *D. miranda*, that the isolation was greatest between races close to each other in their range. King[9] has similar evidence from the *guarani* group. However, even between races of the two species widely separated in origin, the isolation was considerable. One has the suspicion that sexual isolation is common between species which have never had the opportunity to cross-breed, though the evidence is rarely conclusive since seldom if ever do we know the full evolutionary history of the populations.

It is perhaps not surprising that differences in sexual behaviour arose in the experiments involving selection, but were not found in the comparison of populations which had originated independently. If they had occurred in the latter, they could only have appeared by chance, or as a correlated response. It might be expected that random changes in sexual behaviour would be slow, even though they arose as a secondary response to an adaptive change in the population. Mating involves the cooperation of the two sexes and it seems unlikely that a genetic shift in the population causing a change of sexual behaviour in the female, perhaps by a modification of the pattern by which a male recognizes an animal of his own species, would also change male behaviour in a compensating manner (although an exception to this might be in habitat preference). An individual with aberrant sexual behaviour is not likely to leave many progeny. A population gradually changing its genetic situation could only change its pattern of mating by the selection of males capable of responding to the altered female behaviour. This must constitute a brake on the change of mating behaviour either by chance changes in the genetic situation or even as a correlated response to an adaptive change. This will be particularly true of inbred lines in which selection between potential mates is small or non-existent. Reproductive behaviour, excepting perhaps choice of habitat, would therefore be more stable than other physiological systems to genetic changes.

Both the hypotheses that we have discussed demand the development of a previous geographic isolation before sexual isolation can be established. In this sense, the selection hypothesis is perhaps clumsy, since in the formation of a new species showing sexual isolation with the parent species it demands first a geographic isolation and then an overlapping of the species range so that members of the two species can be selected for refusal to crossbreed. It seems to us that sexual isolation instead of being a consequence of geographic isolation, may be a contributory factor in its establishment. The spread of a population into new territory will often involve the occurrence of genotypes with new hereditary habitat preferences. The existence of such preferences amongst *Drosophila* stocks has recently been shown by Waddington, Woolf and Perry [10]. In organisms such as birds, in which rather sudden changes in geographical or ecological range are well-known, learning may play a part, but this may also have an important genetic component. In the genetic constitution of a sub-population which has broken out of the original species boundaries and is spreading into new territory, one must expect to find that a number of adjustments are occurring simultaneously. There is most likely to be, in the first place, an evolution of a new system of habitat-preferences and/or of general activity; in the second, the adaptive characters and general fitness of the migrating group will be attuned to the new circumstances which it has to meet. Both these necessary modifications of the gene pool will be made more easily if the genetic constitution of the sub-population is prevented from continual intermingling with that of the original stay-at-home group. Thus any tendency for preferential mating within the migrating group, and sexual isolation between it and the main population, will acquire selective value. It seems rather probable that a species may be able to spread into new territory even if no sexual isolation develops between the main population and the migrating one; but if the increase in species-range demands considerable adjustment of the genotype to fit the new environment, the evolution of some degree of mating-barrier will undoubtedly be of considerable advantage. Our experiments show that the necessary genetic variability is likely to be present in a population; and the fact that the change of environment involves alterations to the behaviour pattern of the migrating animals makes it more likely that their preferences for sexual partners as well as for habitats, will exhibit evolutionary flexibility. Thus species-spread and sexual isolation will tend to act synergistically.

SUMMARY

1. Partial sexual isolation (between two stocks of *D. melanogaster* differing only in marker genes) has been established by selection of the offspring of flies mating with their own type. This has been demonstrated by a reduction in the number of cross-bred offspring found and also by examination of the progeny of individual females.

2. In a small series of tests, no tendencies towards preferential mating were found to have arisen by chance in a sample of lines which had been inbred or selected for number of abdominal chaetae, although there were differences in intensity of sexual drive.

3. Changes in reproductive behaviour brought about by selection are more likely to affect female than male behaviour. Willingness to cross-breed, in a male, will not materially reduce the number of his pure-bred offspring but in a female it usually will.

4. It is argued that selection pressure against cross-breeding of two

partially separated populations, although probably effective when it occurs, is not likely to be the only mechanism by which sexual isolation between taxonomic groups develops in nature. It is suggested that an important part in the origin of such isolation may be played by the factors (e.g. changes in hereditarily controlled behaviour patterns) which bring about the spread of an initial panmictic population into new geographical or ecological situations.

[1956]

16

THE ORIGIN OF SEXUAL ISOLATION
BETWEEN DIFFERENT LINES WITHIN A SPECIES
(WITH S.KOREF SANTIBANEZ)

If two or more species, sub-species or other taxa are to coexist for some length of time in the same territory, it is necessary both that each group should reproduce itself, and that the genetic systems of the different populations should not become intermingled. The mechanisms by which such a situation is ensured are usually spoken of as 'isolating mechanisms'. One form which they may take, and often do, is the development of mating preferences. It is worth noticing that if these are to operate successfully they must have certain characteristics. For instance, there may be mutations which cause an alteration in sexual behaviour, or in the structure of the copulatory organs, but these would not serve as a basis for an evolutionarily successful isolating mechanism unless they ensure a normal reproductive ability of the mutants. Moreover, when mating preferences occur, they must be complementary. If there are two strains A and B, it is not sufficient that A should mate with A more often than with B unless B also mates with B more often than with A, since otherwise strain B will soon get diluted out of existence.

We still know rather little about the origin of sexual isolating mechanisms. There are two main theories, usually associated with the names of Muller and Dobzhansky. Muller[1] states that the impediments to gene exchange may arise by accident (chance), together with the processes of divergence of the genetic systems. They may be a by-product of other changes. There may be chromosomal genes, arisen by mutation, responsible for the incompatibility, but producing their isolating effect only in combination with other genes. Because of this fact, they 'usually arise in populations separated from one another by outer circumstances. So, it is necessary to conclude, that isolating genes were not established by a process of selection which derived its original advantage from the achievement of the isolating effect itself, but were established, in their original local populations at any rate, through their selective value in other respects, or through drift, as the case may be.'

Dobzhansky postulates an alternative or complementary hypothesis; physiological isolating mechanisms are a product of natural selection. 'Development of all forms of reproductive isolation, including sexual isolation, between populations which have diverged sufficiently to make compromise genotypes unfavourable, may be initiated and furthered by natural selection' [2].

In *Drosophila*, certain strains show mating preferences which might provide a basis for the evolution of sexual isolation mechanisms. Thus Merrell [3] found that certain mutant males (ct, y) mate more often with their own females than with wild type, but males from other strains also prefer these females. Thus these are a one-sided type of mating preference which cannot act as a satisfactory evolutionary isolating mechanism. A rather similar case has been described by Dobzhansky and Mayr [4] who found that in the *D. willistoni* group females from Brazil are more readily inseminated by males from Guatemala and Brazil, the Guatemalan females being less receptive. More interesting is the finding [5, 2] that, in the laboratory, there is greater isolation between populations of sibling species (*D. pseudoobscura-D. persimilis; D. pseudoobscura-D. miranda*) which have been collected from the same locality than between similar populations collected from different parts of the species range. This suggests that the isolation has been developed by the mechanism postulated by Dobzhansky, namely a selection for lack of crossing, occurring of course only in regions where crossing is a practical possibility. But although the facts are consonant with the hypothesis, we do not know exactly what has been going on in Nature, or whether this was in fact the way in which the isolation has arisen.

A few attempts have been made in the laboratory to develop isolating mechanisms between populations. Such experiments have the advantage, over the study of natural populations, that we know in some detail the nature of the evolutionary processes to which the populations have been subjected. They have the disadvantage that sexual vigour often tends to be considerably lower in laboratory strains than in the wild, which suggests that one should bear in mind the possibility that one may be dealing with somewhat atypical animals.

In the laboratory experiments, Koopman [6] was able to intensify the initially fairly low degree of mating preference between *D. persimilis* and *pseudoobscura* by selecting against the tendency to cross-breeding. Knight, Robertson and Waddington[7], by a similar procedure, were able to bring into being, and considerably strengthen, a sexual isolation between flies carrying the mutants *vg* and *e*, which initially showed only a rather slight tendency to homogametic mating.

These two experiments tend to support the hypothesis of Dobzhansky. Knight, Robertson and Waddington also looked for evidence in support of Muller's suggestion, by examining the mating preferences of inbred strains which had been long isolated from one another, but which had never been subjected to any selection against cross-breeding. No such evidence was found, but only a very small number of lines were examined. In view of the great importance of sexual isolating mechanisms for the whole problem of the origin of species, it has been thought worthwhile to study the mating preferences of a number more strains of *D. melanogaster* which have originated in various ways and have been maintained separately for many generations. Any evidence of complementary mating preferences between such lines would give support to Muller's suggestion that such situations may arise by chance.

STOCKS USED

We studied three types of lines, all of which had been maintained as separate stocks for a great number of generations by Clayton and Robertson in this laboratory.

1. *Selection lines.* Eight lines, all originated from Kaduna wild type, which had been selected for the number of chaetae on the fourth and fifth abdominal sternites. Four of them had been selected for a high number of chaetae:

H5/30/R36, selected for 30 generations, relaxed for 36
 (H-5 in table 1)
H1/12/R15, selected for 12 generations, relaxed for 15
 (H-1 in table 1)
H2/20/R46, selected for 20 generations, relaxed for 46
 (H-2 in table 1)
H4/36/R35, selected for 36 generations, relaxed for 35
 (H-4 in table 1)

Four had been selected for a low number of chaetae:

L1/29/R8, selected for 29 generations, relaxed for 8 (L-1 in table 1)
L2/29/R8, selected for 29 generations, relaxed for 8 (L-2 in table 1)
L4/20-1/R47, selected for 20 generations, relaxed for 47
 (L-4 in table 1)
Li/31-3/R35, selected for 31 generations, relaxed for 35
 (Li-1 in table 1)

2. *Inbred lines.* Six lines, also originally from Kaduna wild type, maintained by brother-sister matings for 57 generations. Three of them had been irradiated with X-rays previous to inbreeding: N20-1; N20-9; N20-2.

The other three had not been treated: NC-13; NC-4; NC-11.

3. *Isogenic lines.* Three stocks, originally isogenic, each with a different marker gene: *y* (yellow), *v* (vermillion), *w* (white). One line of each had been inbred, brother to sister for: *y*, 75 generations; *v*, 75 generations; and *w*, 76 generations; while three other lines had been maintained by simple mass cultures for: *y*, 64 generations, *v*, 67 generations; and *w*, 64.

EXPERIMENTAL METHODS

In all cases we used the male choice method designed by Dobzhansky. For the selection and inbred lines, one male was placed in a vial with a female of its own kind and a female of another. The females were made recognizable by a spot of silver paint on the thorax of one of them. In the marked isogenic stocks, one to four males of one kind were placed in one vial with an equal number of females of its own and of another line.

Both males and females were collected within a few hours of hatching, aged in separate vials for six to eight days, etherized and marked at least eight hours before testing. The matings with the selection and inbred lines were made in empty vials. The flies were observed for a period of three to four hours. The moment of initiation and the duration of the first copulation of the male were recorded. After four hours the females were dissected and examined for the presence of sperm in their ventral receptacles or spermathecae.

One line selected for a high number of chaetae was always tested with one with a low number of bristles: L-1 *v.* H-5; L-2 *v.* H-1; L-4 *v.* H-2; Li-1 *v.* H-4; and one of the irradiated lines was always tested to a non-treated line: NC-13 *v.* N20-9; NC-11 *v.* N20-1; NC-4 *v.* N20-2.

For the mutant stocks, males of one line were tested with females of both other lines, maintaining the inbred and the non-inbred as separate groups. The flies were left for 16 to 20 hours at 25°C, in vials with the usual *Drosophila* food medium. After this period the reproductive tract of all females was examined.

With the selection and the non-mutant inbred lines, a certain number of 'female choice' tests were made. The combinations were the same as those mentioned above. One female was placed in an empty vial with a male of her own type, and one of the other line.

Table 1. Selection lines.

1. First Copulation

Male	Homo-gametic	Hetero-gametic	χ^2	Tend.	P.
L-1	12	23	3·4	—	sig.
H-5	25	21	·348	+	—
L-2	28	20	1·33	+	—
H-1	21	23	·091	—	—
L-4	14	26	3·8	—	sig.
H-2	26	19	1·089	+	—
Li-1	14	6	3·2	+	sig.
H-4	12	17	2·79	—	—

2. Male Choice

Male	n	Homo-gametic	Hetero-gametic	χ^2	χ^2 (comb)	Tend.	P.
L-1[1]	25	10	17	4·349	2·174	—	—
L-1	25	14	14	0			
H-5[1]	25	21	13	5·882	1·277	+	—
H-5	25	16	19	·857			
L-2[1]	25	18	14	1·39	1·59	+	—
L-2	25	18	16	·37			
H-1[1]	25	16	13	·74	·199	—	—
H-1	25	14	19	2·23			
L-4[1]	25	9	17	4·701	6·429	—	sig.
L-4	25	9	14	2·013			
H-2[1]	25	18	16	·368	1·116	+	—
H-2	25	20	16	·787			
Li-1[1]	25	5	5	0	1·114	+	—
Li-1	25	11	6	2·228			
H-4[1]	25	9	7	·368	·399	+	—
H-4	25	10	9	·082			

[1] In the upper line of each group, the marked female was of the same strain as the male; in the lower line, the marked female was of the other strain.
— : tendency to heterogametic matings.
+ : tendency to homogametic matings.

3. Female Choice

Female	n	Homo-gametic	Hetero-gametic	χ^2	χ^2 (comb)	Tend.	P.
L-1[2]	25	5	11	2·25	1·125	—	—
L-1	25	8	8	0			
H-5[2]	25	16	3	8·895	16·924	+	sig.
H-5	25	17	4	8·048			
L-2[2]	25	12	5	2·88	7·725	+	sig.
L-2	25	14	5	4·26			
H-1[2]	25	2	12	7·14	3·126	—	sig.
H-1	25	7	4	·82			
L-4[2]	25	8	12	·8	0·318	+	—
L-4	25	12	7	1·316			
H-2[2]	25	9	13	·727	0·845	—	—
H-2	25	9	10	·235			
Li-1[2]	25	6	4	·4	0·2	+	—
Li-1	25	4	4	0			
H-4[2]	25	13	1	10·286	3·468	+	sig.
H-4	25	5	7	·333			

[2] Marked male of same line as the female.

(One of the males was marked with silver paint.) They were observed during three to four hours; the beginning and duration of the copulation was recorded. As during this time the female will only allow one copulation, the flies were discarded after the observation had been completed.

RESULTS

Selection lines (table 1). On observation of the individual matings it could be seen that courtship began almost immediately after the introduction of the flies into the vial. Copulations started within the first two minutes, and most of them lasted for about 30 minutes. As can be seen from table 1, in most cases the first mating occurs more or less at random. The three significant deviations are not very large. In two of the L lines (L-1 and L-4) the males show a slight preference for the H female, while the H male shows random choice. This may indicate a higher receptivity of the H females, a fact also found by Knight, Robertson and Waddington[7]. Only the Li-1 male seems to prefer its own kind. During the four hours for which they have been together, the male may have copulated with another female (section 2). Now, in all except one case, the males have mated with an almost equal number of females of each type. Only the L-4 male maintains his significant preference for the H female. Analyzing the female choices (section 3), there seem to be a few more deviations from the random choice, but one can say that in general the same tendency as in the 'male choice' is maintained. In any case, it seems that there are no consistent data in this group of tests suggesting that any type of preferential mating has become established.

Inbred lines (table 2). The first fact in which these lines differ from the previous ones is in their generally diminished activity. The first copulation is initiated later, about ten minutes after introducing the flies into the vial. It seems also that the total number of matings is less than in the non-inbred lines. There is a much greater correlation between the preference of the male for his first copulation (section 1) and the total number of homo- and heterogametic matings after four hours (section 2).

Table 2. Inbred lines.

1. First copulation

Male	Homo-gametic	Hetero-gametic	χ^2	Tend.	P.
NC-13	22	5	10·704	+	sig.
N20-9	8	8	0		—
NC-11	21	9	4·8	+	—
N20-1	16	9	1·9	+	—
NC-4	28	2	22·53	+	sig.
N20-2	2	12	7·14	—	sig.

2. Male choice

Male	n	Homo-gametic	Hetero-gametic	χ^2	χ^2 (comb)	Tend.	P.
NC-13[1]	25	13	7	3	5·24	+	sig.
NC-13	25	13	5	3·62			
N20-9[1]	25	12	7	2·12	·660	+	—
N20-9	25	7	8	·09			
NC-11[1]	25	10	8	·43	4·406	+	sig.
NC-11	25	14	6	5·33			
N20-1[1]	25	11	9	·33	·166	+	—
N20-1	25	9	9	0			
NC-4[1]	38	24	11	9·133	27·44	+	sig.
NC-4	37	22	4	19·21			
N20-2[1]	38	5	8	·83	5·783	—	sig.
N20-2	37	4	13	6·18			

3. Female choice

Female	n	Homo-gametic	Hetero-gametic	χ^2	χ^2 (comb)	Tend.	P.
NC-13[2]	25	5	1	2·6	2·06	+	—
NC-13	25	4	3	·14			
N20-9[2]	25	11	3	4	·27	+	—
N20-9	25	3	7	1·6			
NC-11[2]	25	5	14	4·26	·948	—	—
NC-11	25	11	8	·474			
N20-1[2]	25	6	2	2	·013	+	—
N20-1	25	5	7	·33			
NC-4[2]	25	8	6	·285	7·99	+	sig.
NC-4	25	12	0	12			
N20-2[2]	25	3	6	1	5·28	—	sig.
N20-2	25	0	5	5			

[1] Marked female of same type as the male.
[2] Marked males of same type as the female.
— : tendency to heterogametic matings.
+ : tendency to homogametic matings.

In the male choice experiments, in all three groups it seems clear that the non-irradiated inbred lines have a definite preference for their own kind, which may be attributable to a slightly greater activity of both unirradiated males and females. The irradiated males seem to mate more at random, although they seem to exhibit a slight tendency towards their own female. When females are allowed to choose between two types of males (section 3), mating is only slightly deviated from random. Only in NC-4 *v.* N20-2 is the preferential tendency significant for both the male and the female choice experiments. The NC females may be more attractive, both to their own and to other males. It is possible that in these inbred lines an incipient type of sexual isolation may have developed.

Isogenic lines (table 3). Both the inbred and the non-inbred lines behave in quite a similar way. After four to six hours very few copulations have taken place, so all matings were left for 16 to 20 hours at 25°C.

The most noticeable fact is the dislike all *non-yellow* females seem to have for *yellow* males, although yellow females accept all males with great intensity. This is a case of 'one-sided preference.'

Of the other two mutants, *vermilion* shows a slight preference for its own females, although this may be a consequence of a higher activity of both males and females.

White females seem to be the least receptive, even towards their own males, which shows a greater tendency towards heterogametic matings.

Table 3. Isogenic lines (male choice)

Inbred

Male	n	Homo-gametic	Hetero-gametic	χ^2	Tend.	P.
y(F75)	100	58	4	68·163	+	sig.
v(F75)	100	93	76	11·032	+	sig.
y(F75)	100	47	0	47	+	sig.
w(F76)	100	33	37	·176	—	—
w(F76)	100	43	60	5·785	—	sig.
v(F75)	100	90	78	5·357	+	sig.

Not Inbred

Male	n	Homo-gametic	Hetero-gametic	χ^2	Tend.	P.
y(F64)	100	62	8	74·89	+	sig.
v(F67)	100	86	84	·157	+	—
y(F64)	100	74	3	106·4	+	sig.
w(F64)	100	65	79	4·861	—	sig.
w(F64)	100	23	51	16·873	—	sig.
v(F67)	100	90	57	32·44	+	sig.

+ : tendency to homogametic matings.
— : tendency to heterogametic matings.

DISCUSSION

In these experiments we have found, with a few of the strains, rather definite evidence of a preference for homogametic matings. The evidence is strongest for the non-mutant inbreds examined. It is a remarkable fact that the three inbred strains which had not been X-rayed before inbreeding NC-13, NC-11 and NC-4, showed the greatest tendency of this kind, while in two of the radiated inbreds N20-9 and N20-1 it was rather feeble. However, in the other irradiated inbred N20-2 the mating preference was probably stronger than in NC-11, and it seems very unlikely that any particular significance should be attached to the radiation which had been administered to some of the lines many generations earlier, at least until many more similar comparisons are available.

Evidence of differential mating was also quite clear for the stocks marked with visible mutants. This is perhaps not very surprising, and the peculiar unattractiveness of yellow males was in fact known previously. A more puzzling, and perhaps more significant fact, is the observation that a tendency towards sexual isolation is much less noticeable among the selected lines than it is among the inbreds. Two factors may contribute to this. In the first place, if, following Muller, we suppose that the mating preferences arise by the chance alteration of the gene pools of the two populations, they will be the more likely to arise the wider the spectrum of genes which are affected. When a stock is selected in two directions, the resulting strains may be expected to differ primarily in those genes which affect the character on which selection was exercised, whereas when a series of inbred lines are prepared from one foundation-stock the genotypic differences which arise by chance should not be confined to any particular category of phenotypic effects. Secondly, during the preparation of a series of inbred lines by setting up pair matings, only those families will survive in which the mating habits of the parents were at least compatible; and this may in effect amount to a certain degree of selection favouring homogametic mating.

In nature, the evolution of sub-specific strains adapted to different environments would probably, in these two respects, resemble the development of inbreds rather than the selected strains of the laboratory. It is likely to be very rare for natural selection to operate on any single phenotypic character of a restricted kind comparable to the 'number of abdominal chaetae' for which selection was made in the stocks tested here; and again in the formation of subpopulations there will be some intra-population selection for sexual compatibility. Thus the fact that tendencies for inter-population isolation have been found in the inbreds, where they must have arisen by chance, seems rather good evidence that the processes suggested by Muller may actually occur in Nature. The major element of doubt in the situation arises from the reduction in general sexual activity which characterizes the inbreds.

The discrimination found was much more marked when males have a choice of two females, than when a female can choose between two males. Merrell[8] states the female choice is more indicative of discriminative behaviour, but, in both types of experiments, it may very well be that it is the female which exercises the discrimination, by either accepting or rejecting the male. The 'male choice' experiment may indicate the greater or lesser receptivity of the female to a particular male, while the 'female choice' method may be indicative of competition between males. More direct observation of behaviour is necessary before one can state in which way the male and the female of two lines may react to one another.

In the lines with mutant markers, the female seems to be the one responsible for a higher or lower discrimination. Several authors have studied the yellow mutation (in *D. subobscura*[9], and in *D. melanogaster* [3, 10]), and they have agreed on the lesser receptivity of non-yellow females for yellow males. In *melanogaster*[10] an important factor seems to be a less stimulating effect of the yellow male; the higher receptivity of the yellow female to both its own and to other males may be an indication of an adaptation of this female to the altered behaviour of the male. In the case of white, the female seems to be less acceptable to any male.

What may be the cause for these mating preferences? It has been postulated that smell may play an important role (certain moths, spiders, cf.[2]) although in *D. pseudoobscura* and *D. persimilis* Mayr and Dobzhansky[11] find that the intensity of sexual isolation is not modified if larvae of both species are grown on the same food. Cutting off antennae also does not seem to alter the response significantly, although it does diminish discrimination of females for winged and wingless males[10].

SUMMARY

1. A search was made for the existence of mating preferences between strains of *Drosophila melanogaster* which had been genetically isolated from one another for many generations, but in which no selection or other procedure which would be expected to influence inter-strain mating had been practised. The mating preferences were investigated by (a) allowing males to choose between females of their own and another strain, and (b) allowing females a similar choice between males.

2. The strains studied were: eight 'selected lines', four of which had been selected for high numbers of abdominal chaetae and four for low numbers; six inbred lines of wild type phenotype; three isogenic stocks carrying the marker genes yellow, vermilion and white.

3. The marked isogenic stocks differed from one another in general sexual vigour, and all non-yellow females show an aversion to yellow males; yellow females are very receptive towards males of any kind.

4. In four out of the six inbred lines there was a statistically significant tendency in the male choice experiments for matings to be homogametic, and in two of these four the same tendency was found in the female choice experiments.

5. Among the selected lines mating seemed to be nearly at random.
6. It is argued that the fact that tendencies to homogametic mating may arise by chance within inbred lines provides rather good support for the suggestion that the initial steps towards sexual isolation between populations in nature may also be sometimes attributable to chance. [1958]

17

BEHAVIOUR, MIMICRY, AND DISRUPTIVE SELECTION

Unpublished correspondence between C. H. Waddington and Professor P. M. Sheppard, University of Liverpool, during November and December 1961.

Waddington to Sheppard, 23 November 1961

Having just received my copy of the new *Advances in Genetics*, I have been reading your article on evolution in Lepidoptera. First of all, I should like to congratulate you on a nice clear account of this really fascinating material. But there is one particular question I want to ask, to which you may have the answer—it may even be in the article, but if so I didn't spot it. A subject which particularly interests me, because it seems to have great 'philosophical' implications, is the relation between the phenotypic appearances of organisms and their behaviour. I was pleased to learn of the case of *Rhododipsa masoni*, which according to the Browers, sits down in the correct orientation to be concealed on some coloured flowers—I hadn't come across this before. But the question is, has this behaviour been brought about by selection of suitable modifiers after the pattern of the moth appeared, or is it something genetically associated with the major gene; in which latter case, it must surely be an expression of some sort of 'intelligence' or adjustment of behaviour in relation to perception by the moth itself. It seems to me that the evidence one wants is: take a recessive factor determining a phenotype whose selective value demands some appropriate behaviour, introduce it into a strain which has not been selected for appropriate behaviour, and maintain it for some time as a heterozygote, in which it is not expressed and no selection for appropriate behaviour goes on; then let it emerge as a phenotype; does the animal behave with sense or not? Could this experiment be done—or has it already been done in Nature—with some of the recessive melanics in moths? *Drosophila* has such boring behaviour from this point of view that I can't think of an experimental set-up with that animal for testing the point.

An incidental, and quite different, point which rather worries me is about 'disruptive selection'. It seems to me necessary to distinguish clearly between disruptive selection with habitat selection and without. It is obvious that if you take two separate lines and select for different things they will diverge; and if you allow some gene-flow between them they may still diverge, depending on such things as the magnitude of the gene-flow, the effective hereditability etc. This seems to have been the situation in most of Thoday's experiments, where he kept an upward and a downward selection line between which his mating system allowed more or less gene flow — in nature this corresponds to two lines favouring two habitats with different selection pressures, there being some gene flow between the two 'ecological races'. But does anything yet show that if a population is subject to two different selection pressures, and yet the whole population mates completely at random in every generation, you achieve a bimodal distribution of phenotypes? This is, I think, the theory that Mather puts forward, and that Thoday suggests his results demonstrate, but which I feel is as yet unproved, and pretty difficult to believe.

Sheppard to Waddington, 27 November 1961
Thank you very much indeed for your most interesting letter and for your kind comments about my article in *Advances in Genetics*. I have not yet been able to look through the volume since my copy only arrived this morning.

With regard to your first query about behaviour and colour pattern, I have no complete answer but can only tell you the situation as I know it. With regard to Browers work on *masoni*, the moth certainly does tend to orientate itself on the flower heads in such a way that its pattern fits the pattern of the flower head. I have seen these moths in the Rocky mountains when the Browers were working on them, and the cryptic coloration is fantastically good. With regard to which came first, the chicken or the egg, I am not certain. The group of moths to which this one belongs are mostly flower sitters and mostly lay their eggs in the flower head, the larvae eating out the newly growing seeds. They are also mostly of bright colours, agreeing in general with the flowers on which they sit. Unfortunately the life history of many of them is so imperfectly known, that a detailed comparison from species to species with colour pattern and behaviour has not been done, although it would be fascinating to do and not difficult. I rather suspect that the colour pattern and behaviour have evolved together, although it is conceivable that the behaviour came first since the position taken up by the moth tends to be that which brings the proboscis over the open flowers of the composite head so that it can feed while sitting on the flower. I think a lot more work will have to be done before one can be at all certain of the evolution of these two complementary characters. I can well see, however, that the two could have evolved together and might well both be multifactorial. With new techniques being developed in the Lepidoptera it might be possible to hybridize closely related species with different patterns and different behaviour and so investigate the genetics of the situation. I think however your suggestion of working with the melanic Lepidoptera is a far better one. I am not yet completely convinced of the difference in behaviour between the Peppered Moth melanics and the typical forms, although it is quite conceivable since I know that many cryptic moths do pick a background which agrees with their pattern to rest on. In fact, most must. Kettlewell's evidence however is not sufficiently strong for me to feel that it would be worth basing too firm conclusions on it. However, if some more work on the subject, which would be easy to do, showed that the melanics tended to sit on black backgrounds, whereas the speckled ones tended to take lighter backgrounds, then it would be easy to investigate the situation genetically, for hybrids can be made between some species as well as between melanics from industrial areas and the typical speckled ancestral type from a place where industrial melanics are not known. In fact, Kettlewell has such hybrids going and I have a stock of them from him where he has crossed melanics from Birmingham with the non-melanic sub-specific form found in Canada which shows no industrial melanism there. He has shown that the dominance of the melanic breaks down in the hybrids, but has not investigated the behaviour as far as I know, though this should be perfectly possible.

The answer then as far as I know it is that the experiment you suggest is perfectly feasible, but has not been done. However, as I say, I think that the suggestion that the melanics and typical forms have a different behaviour pattern in industrial areas needs to be on a firmer foundation that it is at present.

With regard to the matter of disruptive selection I believe that Cyril Clarke and I have evidence that it can work in nature. In fact we have just submitted a paper to *Evolution* on the subject which they have accepted. When I have a spare copy I will send it to you. The situation is this. We have continued work on *Papilio dardanus*, which I mention in the *Advances in Genetics* article, and have shown that the Abyssinian race, where all insects including the mimetic ones are tailed, possesses modifiers which reduce tail length in the mimetic forms but less so in the non-mimetic ones. The rest of the African races as far as we have investigated do not have such modifiers, and the females are everywhere tailless, nor does the Madagascan race in which the females are all tailed but non-mimetic. That is to say, there seems to be a special gene-complex in Abyssinia which we interpret as being due to disruptive selection tending to select for long tails in the male-like non-mimetic females and short or no tails in the mimetic females, the models being tailless. Mating is I think at random, since the males are everywhere the same in appearance, and in butterfly courtship appearance is important. I therefore believe that disruptive selection has selected a special gene-complex in Abyssinia which reduces tail length in the mimetic forms, but less so or not at all in the yellow male-like forms. Of course, the effect can always be fortuitous and not due to selection. This one cannot get over, but the circumstantial evidence suggests that disruptive selection has worked. I think in fact in this case it is not difficult to see how it could act, since you have two major genes, one producing a non-mimetic form, the other producing a mimetic form. All you have to do for disruptive selection to be effective is to select specific modifiers which act in the presence of one gene but not in its absence. Thus, if you selected modifiers reducing tail length, but modifiers that only acted in the homozygote mimetic form, then disruptive selection could be effective, and very effective. I think this may be a slightly different situation to that envisaged by Mather and Thoday, but theoretically not perhaps so different since the modifiers could always act or not act on some environmental stimulus, i.e. modifiers that acted in the presence of one environment but not in another. As I see it, the chief difficulty with disruptive selection is how you keep the two optimal forms you are selecting going in the population. Surely, under most circumstances one will have an advantage over the other, so that the polymorphism, if genetic, will be unstable and disappear as fast as it is created by the disruptive selection. This difficulty of course disappears in mimicry where the advantage of the two optimal types varies with their frequency, and is not maintained by heterozygous advantage or some other such situation.

I am sorry this letter is so rambling, but I am trying to answer you quickly and I will sit down and try and compose a more reasoned argument when I have heard your comments on the present letter, and when I have the manuscript on disruptive selection that we sent to *Evolution* ready to send to you.

Thank you again for your most stimulating letter. I have enjoyed reading it very much indeed—it has set my mind racing.
Waddington to Sheppard, 29 *November* 1961
I was very interested in your letter of the 27th. I am glad to hear that you are thinking of doing something more about the relation between behaviour and colour pattern. I am sure the subject is worth further investigation.

I have several things to say about your discussion of disruptive selection. In the first place, I do not think the problem really arises in any situation where you already have a pre-existing polymorphism in the population. If the population contains two different forms (whether determined genetically or environmentally), then I should have thought it was pretty clear that you could select modifiers which have differential effects on the two forms. Surely this has really been accepted ever since the days of Darwin's sexual selection, where the two sexes are the two polymorphic forms underlying the system. I think your case in *Papilio* in Abyssinia is really parallel to a selection for differential secondary sex characters. It is very interesting to find this operating between polymorphic forms of a single sex instead of between different sexes, but surely the phenomenon is essentially similar.

The real problem about disruptive selection seems to me to arise when you start with a population not containing polymorphism. Consider a uniform population in a region containing two different ecological niches which exert selection pressures for two different sets of characteristics. Will the population become bimodal in phenotype? There are a number of different situations:

1. The initial population becomes effectively divided into two sub-populations by the development of habitat preferences. It seems to me the two sub-populations will then tend to develop different phenotypes, the extent of this divergence depending on the intensity of selection, the degree of habitat preference, the degree of gene flow between the two sub-strains, and so on. This, I think, is the situation investigated by Thoday. In this situation there is some tendency for preference mating between individuals living in the same habitat.

2. There is no tendency to preferential mating, which is quite at random, but there is a correlation between habitat preference and phenotype.

3. There is no preferential mating nor any correlation between habitat choice and phenotype.

It is, in my opinion, only in these two last situations that any serious question about the effectiveness of disruptive selection can arise. My own feeling is that it could only lead to a bimodality in population phenotype if it succeeded in building up something like a super-gene out of a very closely linked set of minor genes. If, on the other hand, all the minor genes under selection continued to recombine fairly freely, I can't see how a bimodality could ever develop.

Sheppard to Waddington, 4 December 1961

I think we are in agreement but talking slightly at cross purposes. Taking an initial non-polymorphic population and your condition 3 (no preferential mating or correlation between habitat choice and phenotype) I believe disruptive selection will normally produce an environmentally controlled polymorphism if it is effective. Thus, any genotype which reacts appropriately in the two distinct environments will be selected, and the environments will come to act as a switch mechanism controlling the morphs. Such a situation is well known in *Papilio* pupae in which there are brown and green forms in many species, the brown forms tending to be found on brown backgrounds and the green on green backgrounds. The switch between brown and green is environmentally controlled and the sensitivity of the insect to the factors tending to produce greenness are affected by the genotype, being different in different races.

I agree with you also in believing that for disruptive selection to be effective in producing a genetic polymorphism a 'major' gene affecting the characters under selection must arise before disruptive selection can produce the bimodal distribution. As I see it, where you have two optimal phenotypes with disadvantageous intermediates you will not get an excess of genotypes tending to be close to each of the optima, when you have a large number of chromosomes and a large number of genes concerned, simply because of the segregation involved. However, directly either a single gene or two closely linked genes, which together have a major effect, occur, then that gene will be selected if it shifts the phenotype sufficiently towards one of the optima. Thereafter, by the incorporation of other linked genes, or by selection of specific modifiers of the major gene, as in the case of the Abyssinian *Papilios* and their tails, the bimodality of the distribution can increase and a true polymorphism evolve. I think that from Thoday's analysis of his selection experiments, when he chose an organism with very few chromosomes and apparently only succeeded in selecting one or two genes affecting bristle number, and closely linked genes at that, in each chromosome, the inference is clear that the bimodality was due to the fact that he was able to pick up 'major' factors rather than true polygenes.

As I see it, in Nature there is another overriding difficulty in producing a genetic polymorphism by disruptive selection, and that is keeping the polymorphism stable. It is easy enough in mimicry where the frequency of the forms determine their selective value, but in other situations there seem to be considerable difficulties. It is true that Howard Levene (*American Naturalist, 87,* 331–3) has suggested that under your condition 3, a polymorphism may be stable even if the heterozygote is not at an advantage, but such situations I feel must be rare. Where they do occur, if they do, I think disruptive selection may be effective, but only by utilizing major genes or short sections of chromosome having a large effect, not by utilization of polygenes scattered all over the chromosomes. However, as I said before, once the major genes are established then modifiers of them can accumulate, and these can be scattered throughout the chromosomes.

Waddington to Sheppard, 5 December 1961
I am glad that our interesting exchange of letters has come to pretty well complete agreement. The only final point I should like to make is that I suspect that the production of an environmental switch mechanism controlling two distinct phenotypes is a relatively rare outcome of disruptive selection in my condition 3. One is always surprised at what genetic variation a population seems to be able to pull out of its hat, but I think it is rather a lot to expect species often to have available the genetic basis on which such an environmentally determined polymorphism can be evolved. It seems more likely that the result will usually be the evolution of habitat preferences and/or differential mating. However, I would agree with you that the outcome you suggest does sometimes occur and presumably has done so in your *Papilios.*

18

A. BEHAVIOUR AS A PRODUCT OF EVOLUTION
Review of *Behaviour and Evolution,* edited by A. Roe and G. G. Simpson
(Yale University Press, 1959).

This book is a very carefully organized cooperative effort, the result of two conferences organized jointly by the American Psychological Association and the Society for the Study of Evolution. The planning began under the chairmanship of Dr Roe in 1953 and a first conference was held in the spring of 1955. This was primarily for the purpose of exploring the possibility of real interdisciplinary collaboration and communication and was successful enough to lead to the holding of a second conference the following year which was organized with the specific purpose of producing a publishable symposium. Drafts of formal papers were circulated in advance to the participants of the second conference where they were discussed in a manner that was essentially editorial. The resulting symposium is, as the introduction points out, a mosaic, but as it goes on: 'A mosaic is (or can be) a picture and not a casual assortment of tiles.' I think most readers will agree that the organizers and editors have in fact been successful in synthesizing a large variety of contributions from different viewpoints into a coherent general scheme. This is a book and not simply a haphazard collection of individual items.

Both the two topics mentioned in the title, behaviour and evolution, are of course enormously complex subjects. No single volume can hope to deal exhaustively with either of them, let alone with both simultaneously. For the purposes of a symposium designed to cover if not all possible aspects of behaviour and evolution, at least a wide and representative range of them, it was necessary to select some focal point around which the various contributions would be integrated. This focus is explained in the introduction in the following words: 'To demonstrate that morphology, physiology and behaviour are aspects of organisms all inseparably in, and explained by, the universal fact of evolution became a principal object of this symposium . . . Behaviour itself, what it is descriptively and how its different aspects may be explained and interrelated through evolution, is the very heart of the book's theme.' The book naturally starts therefore with a general statement of the status of evolutionary theory of the present day. A chapter by Simpson gives a summary statement of the modern neo-Mendelian or 'synthetic theory' and sketches the general principles which have emerged from palaeontological study. Colbert and Romer provide some examples mostly from vertebrates of the conclusions that can be drawn from comparative and anatomical studies as to the relations between evolutionary morphology and behaviour. After this introduction to the basic evolutionary ideas, the second part of the book provides a background for an understanding of the physical basis of behaviour. Beach discusses the evolutionary aspects of psycho-endocrinology. Caspari and Sperry deal respectively with the genetic and developmental basis of behaviour. Pribram and Bullock discuss the evidence from comparative neurology and neuro-physiology.

These two introductory parts bring us to the core of the book, a series of seven chapters entitled 'categories of behaviour'. In the first of these, Nissen discusses the various ways in which behaviour may be classified. For instance, by its functions such as reproduction, dispersal, etc., by descriptive categories such as social behaviour or by the kind of mechanisms involved such as tropisms, conditioning, and so on. He concludes by a suggestion that the behaviour of any animal could be scored according to its effectiveness in six functional categories, namely, sensory capacities, locomotion, manipulation, perception, sensory motor connections and reasoning. He advances the hypothesis that if, at a series of geological epochs, the most advanced, i.e. highest scoring, species at each of these points were taken we would find evidence for some overall consistency in the direction of behavioural evolution. This is, of course, to suggest that the evolutionary process does manifest a general direction of change—a concept which has often been referred to as evolutionary progress, unfashionable though that word may be at this time. In later chapters of this section, some examples of the various categories suggested by Nissen are discussed in more detail. Marston Bates reviews food-getting behaviour mostly in insects and vertebrates. Carpenter deals with territoriality, while Thompson and Emerson discuss social behaviour in general, and evolution of behaviour among social insects. Harlow deals with the evolution of a behavioural mechanism, namely learning. He pays more attention than is usual in such discussions to the learning abilities of quite lowly forms such as flat worms or earth worms, and he argues forcefully that 'there is no evidence that any sharp break ever appeared in the evolutionary development of the learning process'. An article of a rather different kind is that of Hinde and Tinbergen who discuss the comparative study of species specific behaviour and point out 'that the comparative study of behaviour can yield the same type of results as comparative anatomy—a tentative description of the course evolution has taken'. They argue this point on the basis of a detailed study of the behaviour of certain nearly related species, using the concepts and methods of the 'ethology' school. A similar conclusion is reached by Mayer in a chapter entitled 'Behaviour and Systematics' which deals with the subject from a broader comparative point of view. This leads on to two chapters in which behaviour is discussed as a part of the evolutionary mechanisms. Spieth shows that the reproductive isolation between species often involves the behavioural sexual isolating mechanism, while Pittendrigh discusses behaviour as an aspect of adaptation subject like morphological and physiological adaptations to genetic variation and selection.

This brings us to the penultimate section of the book on evolution and human behaviour. Washburn and Avis begin by a comparative review of the behaviour and correlated functional adaptations of monkeys, apes and man. We then pass on to three fascinating and stimulating chapters in which concepts derived from the theory of evolution are applied to higher human cultural behaviour. The subject is approached in turn from a general biological, a psychological and a cultural anthropological point of view. When it is mentioned that the authors are Julian Huxley, Freedman and Roe and Margaret Mead, the reader will easily realize that these chapters, short as they are, are packed too full for it to be possible to indicate the nature of their contents in a short review. The book concludes with an epilogue by G. G. Simpson who makes a manful and indeed remarkable attempt to synthesize the enormously varied contributions of the earlier authors. The success of Simpson's epilogue is in fact a demonstration of the real unity which the book possesses. It is a convincing demonstration that an account can be given of animal behaviour, and at least to some extent of human, in which the theory of evolution provides just as effective an intellectual framework as it can do for a textbook of comparative anatomy.

If a word of criticism were to be offered, it might be this: the relation of the behaviour of an animal to the evolutionary process is not solely that of a product. Behaviour is also one of the factors which determines the magnitude and type of evolutionary pressure to which the animal will be subjected. It is at the same time a producer of evolutionary change as well as a resultant of it, since it is the animal's behaviour which to a considerable extent determines the nature of the environment to which it will submit itself and the character of the selective forces with which it will consent to wrestle. This 'feed-back' or circularity in a relation between an animal and its environment is rather generally neglected in present-day evolutionary theorizing. One might have hoped that it would be more explicitly taken into account in a book concerned primarily with the evolution of behaviour, since it is in relation to behaviour that the circular relation is perhaps most obvious. However, although such considerations are perhaps often just below the surface of the problems discussed by the various authors, they never seem to emerge completely into the light of day. For instance, when Pittendrigh writes that his 'assigned task in this symposium was to discuss behaviour as adaptation', the present writer would have liked to have seen him go on to state that the adaptation must be to circumstances which arise largely as a result of the behaviour. Again, Spieth, in his extremely interesting discussion of the role of behaviour in the reproductive isolation between closely related species, never quite gets around to discussing how far the behaviour itself has played a role in the production of the differentiation between the species. There is here, I think, waiting to be developed a synthesis between evolutionary theory and a consideration of behaviour which goes even deeper than that recorded in this symposium. [1959]

B. LARGE SCALE EVOLUTION
Review of *Evolution above the Species Level*,
by B. Rensch (Methuen, 1959).

This is an English translation of the second edition of Rensch's well-known book published in German as *Probleme der Abstammungslehre* (first edition 1954). The greater part of the book was written during the last years of the war, and the coverage of the literature since then is naturally rather scanty. The theory of evolution, however, is a subject which advances with something of the majestic ponderousness of evolution itself. This is particularly the case when we are considering the major problems of evolution on the grand scale, which is Rensch's theme.

To the English-speaking biologist, Rensch's work will be of particular value in bringing to his attention much important German work which is likely nowadays to be overlooked in the scramble to keep up to date with the most recent advances. For instance, amongst the most frequently cited authors are Abel, Beurlen, Hennig, Plate, Schindewolf, Sewertzoff, and Zeihen. The British reader is likely to miss the names of such palaeontologists as Elles, Swinnerton, Rowe, Spath, Bulman, and others whose contributions were perhaps of little less importance than those of their German colleagues. It is, however, rare enough to find any attempt made to bring the palaeontological findings into relation with those of modern genetics, and this Rensch does with sympathy and critical insight.

The most impressive feature of Rensch's book, and the one which entitles it to be ranked amongst the major works in biology of his time, is the breadth and inclusiveness of the problems which he tackles. After two introductory chapters, on the causative factors of intraspecific evolution and on types of race and species formation occurring in nature, the main body of the book is concerned with trans-specific evolution. Rensch's interpretation of this is, in general terms, that it does not call for the invocation of any special processes other than those which we can see to operate at the intraspecific level. He emphasizes the existence of many systematic regularities and 'rules' which apply rather generally in the trans-specific realm but regards these as high-level resultants of the normal small-scale processes rather than new self-sufficient principles.

In his last chapter, Rensch shows both the boldness and depth of his mind by tackling a problem of which most biologists fight shy —the 'evolution of phenomena of consciousness'. Rensch is one of the few biologists who combine a profound knowledge of the facts of evolution with an intimate working knowledge of experiments on animal learning and capacity for conceptual thought. He has himself (mostly since this book was written) done important work on such subjects as the training of elephants and the (conceivably aesthetic?) painting of chimpanzees. He was also one of the first to construct machines which show the elements of purposive behaviour and learning, in the manner which has since been developed by Grey Walter and others. His views on the relations between evolutionary processes and the phenomena of subjective self-awareness are therefore of the greatest interest. They are both bold and original. He finds himself driven to attribute a capacity for sensation to the lowest organized creatures which can be shown to be capable of learning, that is, coelenterates and possibly even protozoa. He seems, in fact, to agree in general with the outlook of A. N. Whitehead (to whom he does not refer) that something which belongs within the same realm of being as consciousness has to be attributed to all existing things, including the inanimate.   [1960]

C. THE EVOLUTION OF EGGS

Comments on the paper 'The emergence of evolutionary novelties'
by E. Mayr, in *Evolution of Life*, edited by S. Tax
(Chicago University Press, 1959).

In his paper on the emergence of evolutionary novelties Ernst Mayr has given a masterly summary of the way in which neo-Mendelian ideas can provide an explanation for most of the phenomena which have usually been discussed under this head. There is, however, one type of problem with which he did not deal and which I think requires some further discussion. Mayr, like most evolutionary theorists, has been primarily concerned to explain the phenomena which confront comparative anatomists; that is to say, changes in essentially adult structures and functions. The comparative embryologist encounters facts which are in some ways essentially similar to those of comparative anatomy, but it is not quite so clear that he can plausibly appeal to the same processes to explain them. It is a fact which, however, tends to be obscured in some of the simpler accounts of evolution, that the eggs of different classes of animals differ from one another as strikingly and as fundamentally as do the adults. The eggs of segmented worms, for instance, might even be taken to differ from those of insects in a more radical way than do the adult creatures of the two phyla. Less complete but still very considerable differences in egg structure may of course be found within the limits of a single phylum.

These differences must, of course, have been brought about during the course of evolution. It seems to be usually considered that a radical reorganization of the egg will entail a consequential fundamental repatterning of the adult. On this basis it becomes tempting to suggest that in the course of evolution changes in egg structure have occurred and that these have provided the basis for the distinctions between the major classes. Moreover, in many instances it is not at first sight clear what natural selective forces could have been at work in producing the alterations in egg organization, and it therefore becomes tempting to suggest that they may have arisen in a saltatory manner by single mutations. Both these lines of thought seem, for instance, to be implicit in the importance which Dalcq has attached to hereditary changes which, in the process of evolution, have affected egg structures and which he names onto-mutations.

Of course, even the adherents of these views would admit that there are certain alterations in egg structure for which an explanation of the orthodox neo-Mendelian type, depending on the gradual alteration by natural selection of a polygenic system of determinants, can plausibly be offered. For instance, the change from the normal moderately-yolky amphibian egg to the type with a very large quantity of yolk and an extremely rapid development with excessively unequal, almost blastodermic cleavage, in certain frogs living in comparatively dry situations, can easily be envisaged in terms of normal natural selection in favour of fast development, independent of external moisture. Even the very radical change in the early developmental system as between reptiles and mammals can be pictured in a similar way. What is not so easy to see is the natural selective system which might produce, on the one hand, the mosaic organization of the invertebrate spirally-cleaving eggs of Ascidians, and, on the other, the gradient type of organization found, for instance, in echinoderms and, in a somewhat different form, Amphibia. It is, I think, differences such as these that provide the last ditch for believers in a saltational theory of evolution.

I should like to suggest that this difficulty arises in the minds of embryologists primarily because their own work inevitably accustoms them to the notion that the egg comes before the hen. They are impressed by the consideration that an alteration of egg structure will be followed by a re-patterning of the adult. But, it is possible to argue that, from the point of view of evolution, it is the hen which comes before the egg. One may suppose the first step is the production of a radically novel type of adult by any of the numerous mechanisms which Mayr has so well explained. Once natural selection has, for instance, expanded a fold in the integument of a primitive insect into a functioning wing, or elaborated a diverticulum of a fish's intestinal tract into an effective lung, further selection in later generations can set about the task of building up a developmental system which will produce such organs in the most effective manner.

Comparative embryology provides us with many examples of the convergence of developmental systems starting from different initial conditions, but producing rather similar later stages. For instance, somite stage embryos of mammals, reptiles, birds and fish are considerably more alike than are the eggs at the time of fertilization. We may therefore easily conceive that it may not be beyond the powers of natural selection, once a given new adult type of pattern has been evolved, to improve the developmental system by which it is brought into being and in the course of this improvement to alter the organization of the ripe ovum from which the whole process starts. In this context the word 'improvement' might mean an increase in energetic efficiency, an increased canalization against external disturbing environmental circumstances, a greater independence of external supplies of food or moisture, a more rapid attainment of competitive efficiency with those of the other species in the ecological environment, and so on.

So far as I know the most radical morphological alteration of an adult phenotype which has been produced by experimental selection, and which has brought about a general alteration in the gene-pool of a normal wild population, is the production of a polygenic bithorax strain. It is worth noting that this considerable alteration in the pattern of the adult, which if one were slightly lenient one might perhaps allow to be the formation of an evolutionary novelty, does seem to have involved some alteration of egg structure. This is to say, the selected strain includes a maternally-acting recessive gene which causes females homozygous for it to lay eggs which have, in the genetic background provided by the strain, a strong tendency to develop into bithoraxes. This maternally-acting gene must be affecting the structure of the egg in some way, and although we do not yet know what alteration it produces it seems that this could provide a model for the type of evolutionary alteration in egg structure which has been attributed to onto-mutations. The point is that in this case it was only the adults which were selected. The particular treatment, namely exposure to ether vapour, was applied to the eggs during the process of selection in order to persuade them to reveal their hereditary potentialities, but it was not until the eggs of a given generation had reached the adult condition that selection was made for those which would be preserved for further breeding and those which would be killed off. Thus, one may say, selection for a particular type of hen has produced a corresponding type of egg. In so far as this type of argument can be generalized, the brilliant discussion which Mayr has provided of the emergence of evolutionary novelties in adults can be taken also into account for the arising of novel types of egg organization. [1951]

D. THE RESISTANCE TO EVOLUTIONARY CHANGE
Review of *Genetic Homeostasis*, by I. M. Lerner
(Oliver and Boyd, 1954).

The great advances in evolutionary theories which followed the restatement of the principle of natural selection in Mendelian terms can by now be considered classical. A new phase of progress is, however, already under way, and to this Prof. I. M. Lerner makes a stimulating contribution in his new book.

As he points out, 'older evolutionary theories have tended to consider Mendelian populations as relatively uniform, with reserves of variability carried as recessive mutants at individually low frequencies'. But more recently 'the evidence on concealed variability in natural and domestic populations compels us to view Mendelian populations as being exceedingly heterogeneous aggregates of largely heterozygous genotypes'. In fact, we have now come to define the natural sub-units of a widespread species, not in terms of local races of supposedly uniform genetic constitution, but rather as suggested by Dobzhansky, who describes a 'Mendelian population' as 'a reproductive community of sexual and cross-fertilized individuals which share a common gene pool'. This common stock of genes, which is passed around from individual to individual during sexual reproduction, is considered to contain several different allelomorphs at many or most of the possible loci, so that each individual will be multiply heterozygous. Most of the allelomorphs, of course, will produce much slighter developmental effects than those commonly used in laboratory work. If by selection we concentrate the genes acting in a certain direction, and produce a sub-population which differs from the original one by greater development of some character we are interested in (such as a higher milk yield or production of eggs), we almost invariably find that the sub-population has simultaneously become less fit, and would be eliminated by natural selection. It is to this phenomenon that the name 'genetic homeostasis' has been given, implying that if, in a population at equilibrium under natural selection, action is taken (for example, artificial selection) which results in a change in gene frequencies, natural selective processes will be set in motion which tend to restore the gene frequencies to their original value.

The main gist of Lerner's book is a thesis as to the mechanism of these homeostatic processes. One pillar of his argument is the recognition that the effect which a change of gene frequencies has on the incidence of natural selection will depend on the way the genes affect the development of the individual organisms concerned. It is refreshing to find a population geneticist who realizes that he cannot avoid talking embryology and who does his best to do so in a sophisticated rather than an elementary fashion. Lerner's next step is to point out that, in organisms which have been submitted to natural selection, development is to some extent canalized or 'buffered'; that is, it involves feed-back mechanisms of such a kind that the normal optimum adult form tends to be produced under a fairly wide range of conditions. Then follows the really new point which he is making. He argues that this buffering is dependent on heterozygosity, and that when we select for a particular character and thus produce a sub-population with changed gene frequencies, the concomitant lowering of fitness arises because the selection inevitably produces a greater degree of homozygosity, which renders the individuals of the population more at the mercy of chance circumstances during their growth.

There is a certain amount of evidence from several different organisms that cross-bred, heterozygous individuals are better buffered in their development than inbred, relatively homozygous ones; and again, there is some evidence, though not very much, that the more efficient buffering is associated with a greater fitness in the face of natural selection. The evidence does not, however, all tend in one direction; for example, hybrids between local races of *Drosophila* from widely separate localities are less fit than cross-bred individuals from a single population, though they might well be expected to be more heterozygous. It seems probable, indeed, that we have to think of any one natural population as possessing what Dobzhansky has called a 'co-adapted gene complex'; that is, a collection of particular allelomorphs which have been selected because they interact together to give rise to well-buffered development. The most far-reaching arguments in Lerner's book, however, rather discount the importance of the specific nature of the genes which are heterozygous or homozygous. He safeguards himself at various places by stating that the particular character of the genes concerned is, of course, of importance; but he devotes a large part of his exposition to urging that the efficiency of developmental buffering (and therefore the fitness) may depend on the degree of heterozygosity as such, without reference to which loci are heterozygous.

In my view this is throwing the baby (the general principle of the specificity of gene action) out with the bath water (the importance of single major genes as opposed to large numbers of less-effective gene differences). One can, for example, easily obtain populations each of which has its buffering lowered in respect of a particular developmental sequel to a specific environmental stimulus, and one can show that the differences between the populations are genetically determined. Whether this effect on the buffering is due to homozygosity or not, it must at any rate be brought about by specifically different genes in the different populations. Nevertheless, although one feels that Lerner's thesis does not contain the whole truth, it may contain a part of it, and at least point toward some theoretical developments which have not yet been fully worked out. We still understand so little about the mechanisms involved in developmental buffering that one would hesitate to deny that Lerner may be correct in suggesting that there is some non-specific effect of homozygosity on that process, even though one retains the belief that there are also specific effects which he seems to pass over rather too lightly. In any event, one can only be grateful for such a clear and stimulating discussion of topics which are in the very forefront of scientific advance in one of the most fundamental of biological fields. [1955]

E. INDIVIDUAL PARADIGMS AND POPULATION PARADIGMS
Review of *The Case of the Midwife Toad*, by A. Koestler
(Hutchinson, 1971).

It was time, after nearly half a century, for the old 'Kammerer scandal' to be taken out again and examined afresh in the light of our present biology, which has become less passionately concerned with the questions it raises. It is a subject very well suited to Arthur Koestler's combination of a novelist's interest in human motivation with the intellectual acumen of a well-read and diligent nonprofessional student of science. The main part of *The Case of the Midwife Toad* is as fascinating reading as any of the detective stories to which its rather flippant title makes allusion. Some of the appendixes are rather heavy going, but they undoubtedly add to the quite considerable weight of Mr Koestler's argument.

The story is briefly this. In the early years of this century, a brilliant young man, one of the dashing, elegant, music-loving society of Vienna in the last days of the Austrian Empire, began a series of experiments with several species of amphibia—salamanders, toads, and others. Paul Kammerer was fascinated by these animals, and was able to keep living and breeding in captivity species which other zoologists found unacceptably delicate and impossible to maintain.

After several years' work, in which he had reared a succession of generations of some of his animals, Kammerer found himself driven to a conclusion running contrary to a doctrine which was rapidly becoming accepted as orthodox. Kammerer claimed that his evidence showed that if an animal is forced to live under circumstances which are stressful to it, but to which it can make physiological adaptations, and if under these circumstances it succeeds in breeding, then the adaptations will appear more easily in its offspring, and may even appear in them if they are reared in a nonstressful environment.

This is, of course, the doctrine of 'the inheritance of acquired characters', first made fashionable in evolutionary theory by Lamarck in the eighteenth century and accepted for a time by Darwin himself. Since the rediscovery of Mendel's laws of inheritance at the beginning of this century, it rapidly became less acceptable to those who were struggling to lay the foundations of the new science of 'genetics'. By the 1920s, when Kammerer visited London and exhibited his evidence, William Bateson, the main champion of genetics and the inventor of the word, declared it anathema, and almost openly expressed the opinion that Kammerer was a charlatan.

Bateson skilfully and fiercely directed his polemics on to an ever-narrowing front, finally concentrating the whole issue in one experiment, involving the appearance of thickened and darkened lumps of horny skin on the hands of the males of the Midwife Toad. The experiments, which lasted many years, had been completed before the First World War. By the 1920s only one preserved specimen showing the crucial skin-pads still survived, pickled in a jar. The few biologists who tried to repeat the experiment gave up in despair when their animals died or refused to breed.

Bateson concentrated his whole attack on finding fault with this one poor pickled toad, brushing aside Kammerer's statements that he regarded this experiment as rather a side issue, and not his main evidence at all. So the toad was brought out of its jar in 1923 and inspected carefully by the assembled savants in Cambridge and London—but not by Bateson, who thus preserved his freedom of manoeuvre. But if no fault could be found with it in 1923, the situation was changed in 1926. A young American biologist, G. K. Noble, visited Vienna, got permission to inspect the famous toad, and proved conclusively, within minutes of seeing it, that Indian ink had been injected into the places where there were supposed to be naturally darkened pads of skin. Six weeks later, Kammerer shot himself.

So, was Kammerer a faker? It is considered unfair to reveal the secrets of a whodunit, but one may hint at the general character of the sequel by quoting a single sentence: 'Paul Kammerer broke the record by falling in love successively with the five famous Wiesenthal sisters'. There let us leave him under Mr Koestler's quizzical eye.

To the student of science, however, there are some more general questions raised by this story. The issue certainly appeared real enough to the biologists of the 1920s, but would we now take it as such; and if not, how did it come to be accepted? And supposing the issue were real, how does one explain the almost unscrupulous vehemence of Bateson's advocacy? Is such bloody-mindedness endemic to the learned world, as some of the fashionable anti-rational school of today would suggest? (and Mr Koestler sometimes seems to wave a friendly hand in their general direction).

The question whether we would now accept the issues as real breaks down into two questions: Would we be upset if it were proved that the environment in which an animal grew up changed the hereditary endowment which it passed on to its offspring? and: Did Kammerer's experiments look like proving that?

On the first question, the answer is that today we should certainly be very surprised indeed—quite incredulous in fact—if anyone claimed to show that environment could change the basic *qualitative character* of an animal's heredity; but it is realized more and more that there are quantitative aspects which might well be changed in a relatively long-lasting way. Animals, for instance, do not always contain only one representative of each gene for each chromosome, but often have a group of several, or even several tens, of nearly but perhaps not completely identical examples. Again they have some hereditary particles not in the chromosomes in the one and only nucleus of the cell but in other cellular particles, such as mitochondria, of which the cell contains many, again probably not quite identical, representatives.

We know very little about the rates of multiplication of the slightly variant forms of quasi-identical chromosomal or mitochondrial genes, or about the effects on this of demands on the cell's metabolism made by environmental factors. If any awkward facts about 'the inheritance of acquired characters' turned up, there is plenty of room here to hypothesize—and test—theories based on changes in the proportions of closely related genes before we should be forced to the very difficult proposition that alterations had been made in the primary genetic information.

Even so, do Kammerer's experiments look now as though they even began to raise this issue? The answer is, certainly, No. In all his experiments—those with colour patterns in salamanders, as well as those with the toad on which Bateson concentrated his fire—he began with fairly large numbers of animals, of which only a very few left offspring throughout a series of generations; and in all cases, the 'funny appearances' which looked like being 'acquired characters' were appearances which closely related animals were known to be able to produce 'naturally'.

For instance, most toads of the Midwife Toad species do not have the thickened pads of skin, but a very few sometimes do; the ability to produce these pads is already available in the 'gene pool' of the toad population. We have now become so used to thinking of evolution in terms not of individual animals but of whole populations that we should hardly be tempted to ask whether the environment of one individual can affect the heredity endowment of its offspring. We naturally pose the question: does the environment of a parental *population* affect the heredity of its offspring *population*? And, of course, it obviously does, by leading to the *selection* of suitable individual parents to leave more offspring than others. That certainly seems to be what happened in Kammerer's experiments. They all involved very strong selection, which led to 'genetic assimilation', a well-established and quite orthodox phenomenon.

This is a remarkably clear example of what T. S. Kuhn has called a change of 'paradigm'. But why did biologists in the 1920s see evolutionary genetics in terms of individuals, whereas we see it in terms of populations? The answer illuminates the other question of the unusual virulence of the attacks on Kammerer. The emphasis on framing genetic questions in terms of individuals was in fact a reaction against a still earlier paradigm which had stressed populations.

At the time Mendelism was rediscovered, the orthodox theory of inheritance was derived from hereditary processes within populations, deduced by statistical analyses by authors such as Galton and Pearson. When Bateson started talking modern, Mendelian genetics he found himself heavily attacked by a 'population paradigm' orthodoxy. In fact, the attacks became at least as fierce and polemical as anything that Kammerer had to suffer. A whole journal— *Questions of the Day and of the Fray*—was launched by the scientific establishment, and Bateson had his contributions refused publication by the prestigeful periodical *Nature*. It was by such bitter infighting that he eventually established his 'individual paradigm'.

This pre-1914 history, which Mr Koestler only briefly refers to, makes it easier to understand why Bateson in the 1920s had such peculiar ideas of the standards of decent scientific controversy; and also why he was so immovably wedded to his 'individual paradigm' that he could not begin to see that Kammerer's results would have appeared merely interesting, but not so shattering as to be incredible, but only if they were seen from the perspective of a 'population paradigm'. Probably we are being equally blind today to something which will be obvious in twenty years' time—if only we knew, now, what it is. [1971]

19

A NOTE ON EVOLUTION AND CHANGES IN THE QUANTITY OF GENETIC INFORMATION
(WITH R.C.LEWONTIN)

Evolutionary progress has often been discussed in terms of the acquisition of new genetic information. Now the quantity of genetic information contained in a genome can be assessed by the number of nucleotide pairs, i.e. total mass of DNA. There is, in fact, no clear-cut relation between this quantity and evolutionary advance. It is true that metazoa in general contain a considerably larger quantity of DNA than do bacteria. Within the metazoa, however, DNA per cell varies from species to species in a manner which shows little relation to the phylogenetic position of these species. The largest quantities of DNA per genome are found not in the most highly advanced organisms, but, for instance, among vertebrates, in some of the amphibia. There are grounds for thinking that much of this variation may be connected with 'amplification', i.e. the incorporation into the genome of many copies on a single locus, a topic as yet very little understood.

During informal discussions over cocktails at the Villa Serbelloni an argument was advanced by Waddington, and given a more rigorous statement by Lewontin, which seems worth recording (it might be given the name 'The Serbelloni Theorem').

It states that any tendency to increase the quantity of information in the genome during evolution will be held in check because the rate of advance under natural selection will be inversely proportional to the number of informational units.

This was initially advanced on the intuitional grounds that it is always easier to achieve a definite goal when handling a relatively small number of items than when handling a crowd. A more formal argument is as follows:

Suppose the total range possible for a character is from 0 to $2A$ and further suppose that there are n identical gene loci contributing to the character. For simplicity of demonstration let us also assume the genes act additively both between loci and between alleles at each locus. Then each locus has effects 0, a, and $2a$ for the homozygote bb, heterozygote Bb, and homozygote BB respectively. With this notation the additive genetic variance at any locus is $\sigma_{Gi}^2 = a^2 p_i q_i$. But by hypothesis $a = A/n$ so that

$$\sigma_{Gi}^2 = \frac{A^2}{n^2} p_i q_i$$

and the total additive genetic variance over all loci is

$$\sigma_G = \sum_i \frac{A^2}{n^2} p_i q_i = \frac{A^2}{n} (\bar{p}\bar{q} - \sigma_p^2)$$

We see then that for a given total phenotypic range A the additive genetic variance is inversely proportional to the number of loci determining the character. Also, it decreases for a fixed average gene frequency when there is a large variance in gene frequency from locus to locus. Since, by Fisher's Fundamental Theorem of Natural Selection, the rate of advance under selection is equal to the additive variance, our intuitive argument is correct.

Another way to look at it is that any intermediate value of a character can be achieved without additive variance by accumulation of some loci homozygous B and some homozygous b in just the right proportion. If p loci are fixed at b and q loci at B, then $\sigma_p^2 = pq = \bar{p}\bar{q}$ and there is obviously no genetic variance for selection to operate on.

When we turn to the question of the amount of information contained in the genome there is some difficulty of definition. If we regard the genome as a sentence in which the letters are allelic states of each gene, we can use the Shannon definition of information $H_i = -[p_i \ln p_i + q_i \ln q_i]$ where p_i and q_i are the frequencies of the alternate alleles at the ith locus as before. But the product $p_i q_i$ used in the previous formulation of variance is very close to H in form.

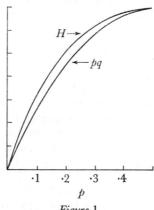

$H \rightarrow$

$\leftarrow pq$

·1   ·2   ·3   ·4

p

Figure 1.

Figure 1 shows the product $p_i q_i$ and H plotted together and rescaled so that they have the same value at $p = 0.5$. Thus we can rewrite the formula for genetic variance as

$$\sigma_{Gi}^2 = \frac{kA^2}{n^2} H_i$$

where k is an arbitrary constant. As a result total genetic variance can be written as

$$\sigma_G^2 = \frac{kA^2}{n} H$$

so that if the average information per gene is fixed the rate of change of the population under natural selection decreases as the information is spread out over more and more genes.　　　　[1968]

DOES EVOLUTION DEPEND ON RANDOM SEARCH?

Most people who consider the theory of biology at the present time seem firmly wedded to the notion that the essential process of evolution is dependent on random search. To give one example: Pattee, in a paper[1], gives one section the heading 'Is evolutionary theory a satisfactory model?' and writes: 'Evolutionary theory is based on the panacean phrase "random search and natural selection in mutable, self-replicating populations" which is used to account, in principle, for the development of almost any degree of molecular intricacy without direct regard for the physical laws which govern the elementary events. It should be borne in mind that the only direct experimental evidence for this evolutionary process is derived entirely from highly evolved organisms of relatively recent and local ancestry.'

As Pattee's sub-heading suggests, many physical scientists find difficulty in accepting random search as an adequate mechanism. The difficulties were strongly voiced at a symposium held at the Wistar Institute in April 1966[2]. For instance, Murray Eden referred to the negligible chance that 'a child arranging at random a printer's supply of letters would compose the first twenty lines of Virgil's Aeneid'. He goes on to offer some numerical estimates of numbers and probabilities considered by him to be relevant to evolution. 'Let us consider first the space of polypeptide chains of length 250 or less. We may think of words which are 250 letters long, constructed from an alphabet of 20 different letters. There are about 20^{250} such words, or about 10^{325}.' He then computes that the total number of protein molecules that have ever existed is of the order of 10^{52}. This is sufficient to indicate the general trend of his argument. Again, Schützenberger writes: 'According to molecular biology, we have a space of objects (genotypes) endowed with nothing more than typographic topology. These objects correspond (by individual development) with the members of a second space having another topology (that of concrete physico-chemical systems in the real world). Neo-Darwinism asserts that it is conceivable without anything further, that selection based upon the structure of the second space brings a statistically adapted drift, when random changes are performed in the first space in accordance with its own structure. We believe that this is not conceivable.'

Before discussing the arguments for and against the adequacy of random search as a mechanism of evolution it is as well to enquire whether in fact modern biology does inevitably come to the conclusion that the directions of evolutionary change are brought about by random search controlled by natural selection.

One argument tending in this direction is a theoretical one. The variations on which natural selections act have their ultimate origin in alterations of the genetic material. These alterations may take the form of changes in the sequence of nucleotides in the DNA, additions or removals of nucleotides, etc. For present purposes we do not need to look any further into the generally accepted principle that these primary genetic changes can be regarded as random mutations. Does it follow that the variations on which natural selection acts are also random? The pebbles forming the gravel on a river bed have their form determined by random processes, i.e. are the results of random search; it does *not* follow that random search plays any important part in the erection of a bridge built of concrete made out of this gravel. The factors that have to be considered in the engineering of the bridge belong, we may say, to a different order of complexity to those involved in the formation of the aggregate of which the concrete is constituted. We need to ask whether the attribution of evolution to natural search does not depend on a similar confusion of different orders of complexity—an error very similar in logical type to Whitehead's well-known 'Fallacy of Misplaced Concreteness'. The question can best be considered in two stages: first, the evolution of higher organisms, as exemplified for instance in the lineages of horses and their relatives, which have been so well studied throughout the Tertiary period; second, evolution at the molecular level, as seen in haemoglobin, insulin, etc., and as, we might argue, it must have proceeded during the origin of life.

In higher organisms it seems fairly clear that the changes which are evolutionarily successful are not in general dependent on single-gene mutations resulting from random mutation. The great majority of random gene mutations which produce effects marked enough to be individually identifiable turn out to be harmful, and to be eliminated by natural selection. There are, of course, some exceptions, such as melanism in Lepidoptera in industrial areas; but even there the incorporation of a major gene mutation into the evolutionary sequence involves a simultaneous selection of a large associated group of 'modifying genes'. In general, however, the evolution of higher organisms depends on the selection of characteristics which are more or less equally influenced by large numbers of genes, i.e. they are comparable to blocks of concrete rather than to individual pebbles. What an evolving population has to do, in fact, is to produce, in some way or other, a phenotype which can find success, in the heterogeneous circumstances of the world around it, in leaving more offspring than other competing phenotypes. As far as its success in natural selection is concerned, it makes no difference whether or not environmental factors operating during early stages of its life history have made a large contribution to the phenotype. In practice, higher organisms have nearly all evolved rather efficient mechanisms for adjusting their phenotypes in a useful manner to surrounding circumstances ('learning' in a very broad sense). Thus physiological and developmental adaptation does usually play a considerable part in determining the phenotype.

In the production of a phenotype efficient at some task important to natural selection, such as, in horses, running fast to escape from enemies, or being able to feed on tough or low-growing herbage, very many genes will be concerned. Going back to the analogy with concrete, one might say that the role of random processes is to produce a gravel which can form a concrete capable of setting into a decent cast of an introduced object, which corresponds to the environment. The cast will not reproduce the details of the object if the pebbles in the concrete are too large, but it will not have adequate strength if they are too small. What is required is some optimum mix. There is nothing but the random process of mutation to produce the pebbles; but this is very far indeed from the conclusion that the casts formed around different objects—i.e. the phenotypes in different environments—are also produced by random search.

It is perhaps worth while to make this point more concretely in terms of a definite example. I will quote one in which I myself have worked with at least some of the genes that might be involved in an evolutionary change. Let us suppose that some environmental change puts a natural selective pressure on a population of *Drosophila* to increase the speed of its flight. Now, simplifying rather drastically, the flight of a fly is brought about by rapid periodic contractions of a series of muscles running between the upper and lower sides of its thorax (figure 1). These change the curvature of the flexible body wall, to which the wing is attached by a complex three-dimensional link which is designed in such a way that, as the body wall changes shape, the wing moves up and down. The forward motion of the insect through the air depends both on the shape of the wing-stroke, which is in some more or less figure-of-eight-shaped path, and on the exact aerodynamic properties of the wing blade. A speeding-up of flight could be achieved by modifying various parts of this mechanism. We know mutant genes, with developmental effects strong enough to make them useful tools in a laboratory, which affect several of the components. Our present knowledge of population genetics would suggest that other weaker-acting alleles of these genes are almost certain to be present in any wild population, and therefore available for natural selection to utilize.

Figure 1. Simplified diagram of the flight apparatus of *Drosophila*.

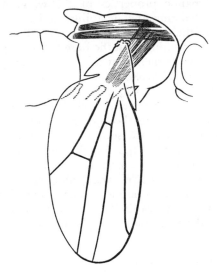

For instance, alteration in the amplitude of the wing-beat or in the exact path followed by the wing-tip might be produced by altering the mechanical or geometrical properties of the surface of the thorax. This can be done by genes such as *dumpy, humpy, thick, thickoid, dachs, dachsous,* etc. Alternatively, the properties of the muscles might be altered, or the supplies of enzymes or substrates which provide the energy resources. Not much is known about genes affecting those parts of the system, but they certainly must exist. Again we know little about genes effecting the wing structure. On the other hand, we know a great many genes affecting wing size, shape and aerodynamic properties. The overall size can, for instance, be altered by genetic systems affecting the number of cells composing the wing, or by different systems affecting the size those cells attain. We know both single genes with strong effects (such as *miniature, dusky,* which reduce the size of the wing cells), and also complex polygenic systems affecting cell size and cell numbers, such as those studied by Forbes Robertson. Again, there are genes affecting the shape of the wing outline. and they can do this in at least two different ways: some, such as *broad* and *narrow,* affect the rates of cell divisions orientated along the length or across the width of the wing. Others, such as *dumpy* and *lanceolate,* affect the manner in which the wing contracts from a fat, inflated condition which it has at one period in its development. Such changes in wing shape certainly affect the aerodynamic properties of the wing. These could also be altered by changing the flexibility of its different regions. The leading edge of the wing is stiffened with a thick vein, and there are four other veins running from the base to the tip of the wing, with two cross-veins between them, but the posterior margin of the wing remains very flexible. Now we know a whole battery of genes, operating in several different ways, which can affect the pattern of this venation and thus the general flexibility of the wing-blade. For instance: *veinlet* removes all the tips of the longitudinal veins; *radius incompletus* removes only the second longitudinal vein; *cubitus interruptus* produces breaks in the fourth and fifth veins; *cross-veinless* removes the posterior cross-vein; and so on. Finally there is another group of genes such as *Curly*, *curved*, and many others which cause the whole wing-blade to be more saucer-shaped, with a concavity either upwards or downwards. All these could affect the aerodynamic properties of the wing and thus the efficiency of the flight mechanism.

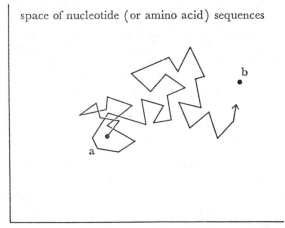

space of nucleotide (or amino acid) sequences

Figure 2. Illustration of a random search process, in which a system is required to move from point a to point b by random changes in single sub-units.

It is from this large array of possibilities that natural selection has to find some solution or other that brings about an increase in the speed of the insect's flight. This is a very different task from that envisaged in the phrase 'random search', which is usually taken to imply that the population has to wait for a new gene to turn up which produces by itself a more or less specific effect which natural selection is demanding. This is perhaps the natural conclusion to draw if one takes the mathematical treatments of Fisher and Haldane as profound contributions to our conceptual understanding of evolution, rather than as drastically simplified first steps towards expressing our theories—which are actually far more complex and subtle—in mathematical terms. A reading of the other great pioneer in the mathematical expression of evolutionary theory, Sewell Wright, would have warned the unsuspecting physicist to tread a little more warily.

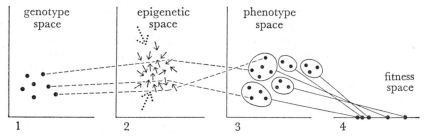

Figure 3. Illustration of the process of natural selection in higher organisms. We start with a population of genotypes in a multidimensional 'genotype space' (1). These are mapped through a multidimensional space of epigenetic operators (2) (operators arising from the environment are suggested by dotted arrows), into an also multidimensional space of phenotypes (3). This is then mapped, by some complex function, into an essentially one-dimensional 'fitness space' (4) in which the only variable is the coefficient of fitness, or Malthusian parameter, i.e. the number of offspring produced. It is in the fitness space that natural selection acts; but fitnesses which are within a tolerance limit of one another are in practice indistinguishable.

If one attempts to diagram the situation in two dimensions, the random search model discussed by Eden might be represented as in figure 2 (the space in which the random walk takes place is of course really multidimensional). The system that really operates is much more like that represented in figure 3. In this we have a multi-dimensional space of genotypes which is mapped through a multi-dimensional space of epigenetic operators into another multi-dimensional space of phenotypes; this in turn is mapped down into a 'fitness space'. This last is essentially one-dimensional, being con-stituted only by quantitative variations in a single fitness coefficient or Malthusian parameter. However, the mapping from the phenotype space into the fitness space is a 'tolerance mapping' in the sense of Zeeman (see, for instance, [4]). That is to say, natural selection cannot effectively distinguish between phenotypes whose fitnesses are too close together (for a more sophisticated discussion on this point see [5]).

When we turn to consider the evolution of molecules, the case against random search is weaker, but by no means negligible. It has, of course, to be quite a different case. We are dealing here with the situation of one gene, one protein, and there is no possibility of arguing that random mutations do no more than produce an assemblage of genes related to the evolutionary product as the pebbles in concrete are related to a bridge. The problem we have to consider is: What are the processes by which evolution produces an enzymatically effective protein? Now the biochemical effectiveness of a protein does not, in general, depend directly on its primary structure, i.e. the sequence of amino acid residues. It depends on the shape and properties of the tertiary, or higher order, configuration into which the amino acid chain becomes coiled. This configuration is determined in large measure by the primary structure, but the general recognition, which is becoming more and more widely accepted, that proteins may exhibit allosteric properties in relation to other molecules in their environment, shows that the primary structure does not entirely determine biochemical properties of the protein, without reference to anything else. Thus even a single protein molecule has a 'phenotype', in the characterization of which the environment plays a part.

The effectiveness (=natural selective value) of a protein depends mainly on one or a few reactive sites on the surface of the tertiary configuration into which the protein becomes folded. These sites involve only a small fraction of the whole amino acid chain. The rest can be regarded as packing or scaffolding, and have not much more to do with the activity of the protein than had the scaffolding which allowed Leonardo da Vinci to lie on his back in a convenient position to paint the ceiling of the Sistine Chapel to do with the resulting art-work.

The sequence of amino acids in a protein can be altered not only by changing one nucleotide pair in the DNA—which is presumably a completely random process. There are also possibilities of re-arranging, duplicating, or inverting whole sequences of amino-acids, comprising considerable numbers of units, by processes such as intra-cistronic recombination (cf. the ideas of Smithies on the determination of antibodies). These processes can also be considered 'random', but in a sense quite different from that which would be implied if one was thinking only of a 'random walk' through the space of all possible sequences of an amino acid chain of a given length. Moreover, as Fraser pointed out at the same Wistar Symposium[6], as soon as organisms reach an evolutionary level at which they have diploid genomes, the possibilities of recombination enormously increase the efficiency with which random processes can explore a space of possible variations. Utilizing this, he has written computer programmes which lead him to assert that 'the genetic mechanism is a superb learning device', in contrast to the biologically less sophisticated computerers, such as Murray Eden, who conclude that 'every attempt to provide for "computer" learning by random variation in some aspect of the programme and by selection has been spectacularly unsuccessful'[7].

The fundamental problems of molecular evolution seem to me to be these:

1. We have somehow to get a protein which folds up into a tertiary structure which possesses a site which to some extent 'fits', and therefore has enzymatic activity on, a certain substrate. The properties of this site may be influenced not only by the primary amino acid sequence, but also by other molecules in the environment, among which the substrate molecules themselves may be important.

2. The next steps in evolution involve improving and stabilizing the effective characteristics of these sites.

For process 2 evolution could pursue either of the two courses which are open to it in the evolution of higher and more complex entities; it could either improve the adaptibility of the molecule, i.e. the ease with which the properties of the reactive site could be altered by the substrate itself, so as to render the site more effective; or, secondly, it could pursue the strategy of canalization, i.e. elaborate the rest of the amino acid chain in such a way that it stabilizes an effective reactive site, independent of the presence of the substrate molecule. To perform these tasks there are available processes which are random at two levels; firstly, random changes of single nucleotides, and secondly, rearrangements of sequences of nucleotides.

For the first of the evolutionary steps—getting a protein molecule which has *some* biochemical activity of a useful kind—the same two types of mutational process are available, but I have the feeling that the former must be more important in this connection.

Until we know more about the nature of active sites on enzymes, how detailed their specification has to be, and how much they can exhibit adaptation to the substrate molecules, it seems impossible to estimate whether these random processes would have a reasonable chance, firstly, to bring effective protein molecules into being, and secondly to improve and stabilize them into the types of proteins which we meet at the present day. It is, however, only in this area that the adequacy of random search as a basic evolutionary mechanism presents a problem. As argued in the first section of these notes, we certainly do not have to suppose that a vertebrate eye, the leg of a horse, or the neck of a giraffe is in any important sense the result of random search.   [1968]

21

COLONIZING SPECIES

The purpose of the symposium is to discuss, in terms of evolutionary theory, the situations which have arisen from the introduction of plants and animals from one part of the world to another. These transplantations have, in effect, been a series of experiments in evolution. As such, they are potentially much more informative than most laboratory experimental work, since they have faced the introduced species, not with some simple defined change in selective conditions, but with a whole new ecological system in which the species has to find a place for itself.

Evolutionary theory is at present attempting to move on beyond the extreme simplification of outlook which was necessary in the first stages of the development of a genetically based theory. In particular, it seems desirable to develop two rather more subtle lines of thought:

1. The natural selection which guides evolutionary change acts primarily on phenotypes, and only secondarily on genotypes. Therefore, any agents which alter or limit phenotypes, e.g., physiological adaptation, developmental canalization, etc., may have evolutionary consequences (see discussion, in [1]). Are any cases known in which selection of an introduced species for some special physiological adaptation (to temperature, reversal of seasons, etc.) seems to have had important consequences (which might amount to genetic assimilation)?

2. The natural selective pressures acting on an organism are not simply given by the circumstances, but depend to some extent (particularly in animals) on the activities of the organisms themselves, i.e. natural selection forms part of a 'feedback' loop, with 'position in the ecological system' as the other link. An organism must be regarded as something with many potentialities, that has to

make its way by fitting, in some way or another, into an ecological system composed of other pluripotential organisms and a diverse collection of available physical factors. There may be a number of different ways in which a given species, introduced into a foreign ecological system, could earn a living within it. We have a situation comparable to some of those investigated in the theory of games. Do we find any cases in which an introduced species has fitted itself into a rather different ecological niche than the one it occupied at home, e.g. feeding on a different species of plant or adopting a different mode of life, from arboreal to ground level? Presumably once any new evolutionary path begins to be followed successfully, the 'inertia' produced by selection of many genes will tend to direct further advance along the same line.

In the 'evolution game' which it is playing, a species has to contend with unforeseen eventualities which the future may bring— a new parasite, a new predator, possibly an Ice Age. Another element of uncertainty arises from the fact that there may be several different ways in which the species makes a place for itself within the whole ecological network available for its exploitation— it could change its food habits or length of life cycle, or it could migrate to another locality, and so on. The game it is playing is perhaps best formulated as a zero-sum three-person game, the players being (a) the species or population under consideration, (b) the whole environment, organic as well as inorganic, that impinges on that species, and (c) the bio-system that would occupy the living space of first player, if it were eliminated. This third player may include species which are also involved in the second player, but by formulating the game in this way, the third player is reduced to a dummy whose only function is to absorb the gains and losses of the first player; in this way we retain the advantages of dealing with a zero-sum game and only have to consider the moves of two players, the species and Nature.

If, for simplicity, we regard the generations as distinct, then each generation is one game, and the 'score' is the number of offspring which reach reproductive age in the next generation. There are several types of 'moves' which the population can make, but they are all constrained by the consideration that changes in the assemblage of genotypes in the population can be produced only by mutation, hybridization, or natural selection. A population has, of course, no intelligence of its own which would make it possible for it to choose which move to make, i.e. to adopt a strategy. But no large population is fully panmictic; it is always broken up, if only by distance, into a number of smaller sub-populations which are partially independent in genotype. Each sub-population will make a somewhat different move, some of which will be more successful and others less; a global or 'Monte Carlo' strategy will emerge as that sequence of moves that has proved most successful up till the stage the game has reached. As we shall see, there are really many different games going on simultaneously, affecting different levels of individual and population organization, and each game elicits a corresponding strategy.

What global strategies can or will be adopted under various circumstances? One strategic question has been discussed by Lewontin[2]. As he points out, the theory of games is essentially a theory of how to play safe; its basic notion is the Minimax strategy, i.e. the strategy which maximizes the minimum gain the player will get if his opponent does his worst. On the other hand, the rules for each individual play of the game are those of natural selection; and this is a Maximax rule, i.e. it simply reinforces the maximum gain. In the actual series of games which a given species may play, its opponent, Nature, may not, in fact, confront it with the most unfavourable circumstances imaginable. So long as this is the case, the species can 'get away with' a variety of strategies, and the question of modifying its genetic system so as to produce an optimal strategy is irrelevant. It seems likely that a great many, perhaps even the majority of, biological species are in this condition. It is difficult to believe that plants which have adopted an apomictic system or lower organisms which rely on parasexual reproductive mechanisms, such as mitotic recombination, are employing optimum evolutionary strategies; they are probably simply persisting with strategies that have been adequate so far against a lenient opponent. The situations which are theoretically interesting, however, are those in which there is some pressure to alter the strategy.

As a simple model system consider a population with annual generations, subjected to environmental conditions which vary from year to year. In year i, the available genotypes G_i will develop, under the existing environmental conditions E_i into a set of phenotypes P_i. Each of these will have a selective value S_iP_a, S_iP_b, S_iP_c, etc., which will depend also on the environmental conditions E_i. The result of the selection will be the survival of a set of selected genotypes Gs_i. It is from these that the genotypes of the next generation G_{i+1} must be produced, which, developing in environment E_{i+1}, will give rise to the array of phenotypes P_{i+1}. The necessity for survival of the species is that this array of phenotypes contains individuals that succeed in leaving offspring in the prevailing environmental conditions.

There are quite a wide variety of 'strategies' by which a species can attempt to meet this requirement; in fact, they affect different levels of biological organization, so the species is playing several different games simultaneously. Some examples of possible different strategies are as follows:

1. At a fundamental biochemical level, there are alternative strategies possible in the organization of the genetic control of enzyme sequences. Consider a metabolic pathway in which successive steps are calatyzed by enzymes P, Q, R, S, T, ... As Kacser[3] has pointed out, in some cases it is found that one of these enzymes, say R, is rate-determining for the whole sequence of steps, the throughput being highly dependent on the quantity or activity of R, but little affected even by considerable changes in the activities of the other enzymes; in other metabolic pathways, all of the enzymes may have more or less equal importance in controlling the over-all flow through the system. If the first strategy is adopted, the system is little affected by mutations or environmental effects controlling the nonrate-determining enzymes, but it is very sensitive to effects on R; with the second strategy, the system is affected somewhat by mutations or other influences on any of the enzyme proteins, but is not affected drastically by any of them. The second would therefore seem to be the Minimax strategy, but a species might often be able to get away with the first gambit, in which it would only rarely suffer any loss of efficiency, at the expense of failing completely in a few individuals.

2. In order to meet the demands of differing environmental effects on development, and on selective pressures, a species has, in general, to preserve considerable genetic variation within its populations. But this variation can be deployed in a number of ways: (a) The species can become very good at producing one particular phenotype under almost any circumstances, relying upon the environment always offering a possibility for this phenotype to get by. This leads to the evolution of systems of developmental canalization of the phenotype. (b) The species can become good at doing one or another of a few alternative things. This leads to switch mechanisms between canalized phenotypes, e.g., in species which have hot and cold weather or aquatic and terrestrial forms. (c) The species can allow the environment to have a strong influence on individual ontogeny, provided it is ensured that the environmental modifications are toward the selection optimum for that particular environment.

This leads to the evolution of developmental systems which are highly adaptable. (d) The species can have a development which is relatively unaffected by normal environmental variations, but in which most genetic changes come to phenotypic expression, and can rely on its wealth of genetic variation always to throw up some phenotypes near the selection optimum. This leads to systems in which there are considerable random or periodic changes in the gene pool from time to time, but little general long-term movement in any particular direction (e.g. fluctuations in inversion frequencies according to season, as in some species of *Drosophila*).

3. A major strategic alternative for a population is whether it continues with the degree of genotypic continuity which entitles it to be called one species or whether it becomes subdivided into recognizably different taxa.

4. A colonizing species, i.e. one which finds itself in a new locality, has another type of game in front of it, namely to decide into which of the available ecological niches it will fit itself. A similar problem of course also arises, in a less extreme form, for stay-at-home species which find that the environment has changed around them. What are the evolutionary consequences on town sparrows of the disappearance of horse manure from our streets; or from the discovery by blue tits that milk bottles on doorsteps offer an attractive source of a type of nourishment they had never indulged in before?

It seems to me that what is needed, at the present stage of our understanding of evolution, is not so much a greater elaboration of formal theories of quantitative and population genetics of the kind we have been producing for the last twenty or thirty years, or even more analyses of wild populations in terms of genetics divorced from their ecology. What we need is more knowledge about the ways in which populations, in fact, meet evolutionary challenges: what intensities of natural selection can they put up with, how far and how fast can they modify their phenotype (including their habits)? Colonizing species are ones which we know to have been confronted by a challenge—that of their new location; and we often know, even quite precisely, how long they have been facing it. We know that the red deer of New Zealand, derived from Scotland, live under somewhat different conditions from those of their ancestors, eat rather different food, and look somewhat different; but we have, I think, almost no idea how far they have been altered genetically. Again, as F. Wilson has pointed out, the mosquito fish *Gambusia* has been introduced into many parts of the world, including some which would appear to be colder than the species might be expected to tolerate. How much change has occurred in the genetic system controlling cold tolerance, and has this taken the form of a lowering of the whole tolerated range, including the upper end, or of a broadening of the range, so that the lower level drops leaving the upper level unchanged? It is facts of this kind that the evolution theory now needs. But in order to get them, and to understand them, we shall also have to work out in greater detail a more flexible but also more down-to-earth and less abstract body of theory, perhaps somewhat along the lines sketched above.   [1965]

22

THE PRINCIPLE OF ARCHETYPES IN EVOLUTION

Classical neo-Darwinian theory may be summarized as a system which involves: (a) random gene mutation, treated as a repetitive process so that each mutational change can be assigned a definite frequency. (b) Selection by 'Malthusian parameters', i.e. effective reproduction rates. (c) An environment which is treated as uniform, that is, it can be neglected. (d) The phenotype has no importance other than as the channel by which selection gets at the genotype.

This system is theoretically a closed one, which does not lead to continued evolution, but at best to a passage leading to a state of equilibrium. The possibility of continued evolution requires the postulation of one or more of the following additional points: (1) A continued change in the environment, arising independently of the existence of the organisms within it. (2) An initially heterogeneous environment, whose heterogeneity is continually increased by the fact that different populations evolve into adaptation to the initial sub-environments; i.e. the organisms adapted to other environments form part—a changing part—of the environment of the organisms in the sub-environment under consideration. (3) The existence of epigenetic organization of the phenotype, so that the phenotypic (i.e. selectional) effect of a gene mutation is changed when it occurs in a later-evolved organism from what it had been at an earlier stage. (4) The possibility of the occurrence, at later stages, of types of gene mutation which were theoretically impossible at earlier stages (e.g. the occurrence of an intra-genic duplication would then make possible types of mutation which previously could not occur.).

It seems almost certain that all these factors have in fact played a part in evolution as it has actually occurred. The questions then to be asked are: how have they operated so as to result in the appearance

of only a restricted number of different types of organisms? Why
are there no organisms which simultaneously exploit the full
potentialities of myosin (active movement) and chlorophyll (photo-
synthesis)? Why are there no vertebrates as thoroughly hermaph-
roditic as some molluscs? In other words, once you have got a theory
which makes continued evolution possible the major problem is to
provide a general theory of phenotypes. Factors 1 and 4 above clearly
do not provide any basis for limiting the freedom of phenotypic
forms. The choice lies between factors 2 and 3; and it seems doubtful
whether the types of environment (i.e. of selective pressure) which
might arise (factor 2) could be shown to provide any principles of
limitation. It therefore appears to me that the crucial issue is factor
3, i.e. the epigenetic organization of the processes by which the
genotype becomes developed into the phenotype.

The main elements in the theory of phenotypes are: (a) 'The
canalization of development' (i.e. what is often crudely called
"developmental homeostasis", *plus* switching mechanisms); (b) the
heritability of developmental responses to environmental stimuli;
(c) the theory of 'archetypes'.

Items (a) and (b), taken together, produce, among other things,
the process of 'genetic assimilation'—the only radically new,
experimentally verified, evolutionary process discovered in the last
twenty years (?). It is item (c) which I wish to discuss further.

The Principle of Archetypes in Evolution

Evolution is brought about by the natural selection of random
variations which occur in the genetic material, which—in conjunction with environmental influences—determine the phenotypes of
the organisms which have to find some way of earning a living in the
natural world. The point towards which attention is being directed
in this Note is the fact that the individuals on which natural selection
operates are *organisms*. That is to say, their character is not a mere
summation of a series of independent processes set going by a number
of disconnected genetic factors, but is instead the resultant of the
interaction (involving all sorts of feedback loops, mutual interference, mutual competition, etc.) of a number of elementary
processes for which the individual genetic factors are responsible.
Now a major characteristic of such interacting systems is the existence of 'threshold phenomena'. Suppose that we have an organized
system subject to variation and natural selection makes the demand
on it that a certain parameter should reach a minimum value p. It
will very likely be the case that the first variation which attains this
value p actually interlocks with the other elements in the organized

system so that, in fact, it attains a value of $p \times n$, where n may be quite a large number.

Consider an excessively simple case. Imagine in a two-dimensional universe an organism which was faced with the necessity to protect its internal contents against the random buffetings of the external world; the organism is to be thought of as having protective elements in the form of three rods of relatively rigid material joined together at the ends by joints around which movement is possible. Now if these three rods are such that the sum of the lengths of the two shorter elements is less than the length of the longer element, they can form a sort of protective arc whose strength can only be in-creased by increasing the friction at the terminal joints. This will give the whole configuration a certain, but quite restricted, resistance to deformation. If, however, the length of the two shorter arms is increased so that the whole system can join together to form a triangle, the overall resistance to deformation of the configuration suddenly leaps up by many orders of magnitude—probably by much more than existing circumstances were demanding of it.

I should like to suggest that a major factor in large-scale evolution has been this sort of 'hitting-the-jackpot'. For some relatively trivial local reasons certain arthropods, varying indefinitely over a whole range of phenotypes under the influence of natural selection, come up with a form involving three pairs of legs, two pairs of wings and respiration by tracheae. It turns out to their surprise (if a phylum can feel such a sentiment!) that this particular pattern of organiza-tion opens out the whole range of the insects. Another group comes up with an 'archetype' of eight legs with no proper division between the head and the thorax, etc., and they find they can develop into the whole range of the Arachnida.

There should be—but unfortunately there are not yet—large chapters in the mathematical theory of evolution concerned with such questions as:

1. Under what circumstances does selection for a slight increase in a parameter x in an organized system lead to an increase in x by several orders of magnitude?

2. How can you distinguish between topologically distinct patterns of organization? Consider automobiles. There is the motor bicycle pattern, with one wheel to drive, one to steer and no lateral stability. There is next the three-wheeler, with either one to drive and two to steer or *vice versa*, and some lateral stability. There is the standard four-wheeler, usually with two to drive and two to steer, but possibly with both pairs driving and/or both pairs steering. Then there is the

tractor vehicle, which amounts to an infinite number of driving units and I am not sure how many steering units. Is it conceivable that there could be a biological topology which would show why you can get large classes of animals with two, three or four pairs of limbs, or an indefinite number of pairs as in Myriopods, but there is no major class with six or seven pairs? Again, you can get symmetries based on two, four, five and six axes but not, apparently, on seven. Or is this perhaps merely a contingent accident that no animal phylum during evolution happens to have hit on a seven ray symmetry? But even without tackling the possibly insoluble problem of why evolution hasn't done what it hasn't done, we certainly cannot be content without some further understanding of how these small-scale processes of natural selection of minor variants in relation to immediate needs have produced a restricted number of basic 'archetypes'—the protozoan, the annelid, the insect, the vertebrate—which are flexible enough to become adapted to almost any system of life yet have sufficient inherent stability to do so without losing their essential character.

3. The notion of an archetype, as described above, has been made too 'simpliciste' for purposes of preliminary exposition. What we really have to do with always involves time, and in the context of evolution, involves it at two (or more) importantly different scales. In the first place, an archetypal form of individual organization (an insect, an arachnid, etc.) is not simply the conventionally accepted adult configuration—it is the whole epigenetic trajectory leading from the egg to the adult. It is an archetypal chreod.

Here it is worth noticing the difference between an archetypal and the usual kind of epigenetic chreod; both are 'canalized' and protected by threshold-like barriers against disturbances; but in normal chreods the thresholds lie where they do only because natural selection has put them there, by selecting certain values of particular parameters. In an archetypal chreod on the other hand, the position of the thresholds is fixed by internal necessity. (Is this what Rene Thom calls 'stabilité structurelle'?). You can't have a triangle unless $a+b>c$. Of course it is natural selection which picks out one archetypal chreod as something to be exploited, but it has not created it; whereas it has created the normal epigenetic chreod out of a situation in which no chreod is logically necessary.

To return to the main argument, an archetype should probably—I am not quite sure of this—be regarded as time-extended on the evolutionary time scale. You don't just get a 'horse archetype', a 'dipteran archetype', but you get a 'horse family archetype', with inbuilt characteristics of directions in which evolutionary change can easily go. [1967]

23

THE EVOLUTIONARY PROCESS

It is essential that biologists should be as careful and precise as possible in formulating an adequate and profound theory of evolution, since it is now widely recognized that this is the most central theory in the whole of biology. In the first quarter of this century the most influential biologists considered that the basic characteristic of life is to be sought in the metabolic activities of living things. A biological system, they pointed out, is one which takes in relatively simple substances from its surroundings and elaborates these into more complex substances. It appears to operate in a manner directly opposed to the second law of thermodynamics. Later it was realized that these apparent contradictions of fundamental physical law are only local and temporary, and, indeed, occur only in parts of the total system and not in the complete system which comprises both the living organism and its environment. The search for a fundamental theory of biology then shifted towards genetics. A living system began to be regarded as basically one in which there is hereditary transmission of mutable information. This makes possible, or indeed inevitable, the process of natural selection, and thus of an evolutionary process which can be held responsible for all the further elaboration which we find in the living world today.

This Neo-Darwinist view of the basic nature of life is the dominant one at the present time. It is the view most generally discussed by the people interested in scientific theory, most of whom approach the subject primarily from the side of physics. In my opinion, however, it is inadequate in several ways, and I should like to discuss some of these.

THE NECESSITY OF THE PHENOTYPE

The term 'Neo-Darwinism' is often applied very loosely, to include almost any recent evolutionary theorizing which has been influenced by Mendelian genetics. Many of the successes claimed for it by its defenders (e.g. the understanding of evolution in social insects) should, in fact, be attributed simply to Mendelism itself rather than to Neo-Darwinism. In its strict sense, it refers to the mathematical models developed originally by Sewall Wright, Haldane, and Fisher, and elaborated since then by many later authors. This strict Neo-Darwinism does not involve any necessity to refer to the phenotype. It postulates a population whose individuals possess genotypes selected out of a collection of genetic units (genes, cistrons) G_1, $G_2, \ldots G_n$. These are to be replicated by some process, which produces a new genotype, consisting of units G'_1, $G'_2, \ldots G'_n$. It is supposed that if one unit, G_j, undergoes a change (mutation) into a new form $G_{j'}$, this alteration will be copied by the replication process so as to produce a new unit $G'_{j'}$. In order to bring about a process of evolution, it is necessary to make one further assumption, namely that there are interactions with the environment which bring about differences in the frequencies with which G_j and $G_{j'}$ are represented in later generations. These interactions constitute the process of natural selection. Provided that this occurs, and that there is transmission of mutable hereditary information, then evolutionary changes, in the sense of alterations in the frequencies of G_j and $G_{j'}$ in the population, are inevitable.

This statement makes no reference to anything which needs to be considered a phenotype as opposed to the genotype. Neo-Darwinist evolution is, therefore, theoretically possible in systems in which no such distinction is called for.

Now, in practice we do not actually find any such systems in the biologically evolving world, which relies mainly, if not exclusively, on a carrier of hereditary information, DNA, which can be replicated only with the aid of a phenotype consisting of enzymes, such as polymerases, and those in the biosynthetic pathways producing necessary subunits (nucleotides, sugars, ATP, etc.). There are indeed systems which can be thought of as exhibiting hereditary transmission of mutable information without the participation of a phenotype, but they are normally classified as belonging to the inorganic world. An example, discussed by Cairns Smith[1], is the transmission of imperfections of the lattice during the growth of a crystal. He has suggested that a process akin to natural selection would occur in such situations, and that there might be an effective 'pre-biotic evolution'. It is, however, by no means clear how this 'natural selection' could amount to anything more than differences in the reliability with which the imperfection is transmitted; and this would be dependent on the nature of the imperfection itself, not necessarily involving any interaction between the hereditary information and the environment. Such systems do not, for this reason, provide true models of biological evolution for which it is essential that it be the environment (i.e. something outside the organism) which exerts natural selection by determining the frequencies with which alternative hereditary units are represented in later generations.

Systems in Neo-Darwinist evolution must find some way of reconciling two rather conflicting requirements: (a) they must have a method of storing genetic information in a form which is sufficiently unresponsive to environmental influences to be reliable, and (b) they must interact with the environment sufficiently to feel the effects of environmentally-directed natural selection.

Any system which incorporated both these requirements into a single substance, which acted both as memory-store and as environment-reacter, would almost certainly have to exhibit Lamarckian effects in which the environment could directly produce changes in the content of the stored genetic information. It is, perhaps, not necessary to discuss further how far it might be theoretically possible to design such a system, since we know that this is not the method that the living world has adopted. It has followed the strategy of storing genetic information in a chemically unreactive substance, DNA, which has minimal interactions with the environment, and developing from this (through the intermediate step of RNA) a phenotype based upon the highly reactive proteins, which both exert influences on and are influenced by the environment.

Even if ways can be found around the conclusion that the existence of phenotypes in systems undergoing evolution is a logical necessity, it would remain true that it is such a general characteristic of the evolutionary process actually employed by the biological world that no theory which omits it can be considered adequate. The model of evolution described in the first paragraph of this section, which was used for the formulation of the mathematical treatments of Wright, Haldane, and Fisher, suffers from this deficiency.

The introduction of phenotypes into this system of ideas involves a rather radical re-casting. It is sometimes considered enough merely to insert into the old equations a 'fudge factor', the 'heritability', to allow for the fact that the phenotypes on which natural selection acts give only partial information about the genotypes. But this is not sufficient to encompass the necessary complexity of the situation. A phenotype is reactive to the environment in several different ways. It is not only responsive to the fundamental process of natural selection (i.e. to influences on the rate at which it transmits its genotype to later generations), but it also submits to actions which modify the epigenetic processes by which the character of the phenotype becomes determined during development, and it may also act positively on the environment so as to alter it.

The last of these, the influence of the phenotype on the environment, can perhaps be neglected in a treatment of general principles, since it is of importance mainly in the later stages of evolution, when organisms have attained some complexity. But the modifications of the character of phenotypes by environmental effects cannot be left out of account, since perhaps the major problem of the whole of evolutionary theory is to account for the adaptation of phenotypes to environments.

THE NECESSITY TO CONSIDER MORE
THAN ONE ENVIRONMENT

The strict Neo-Darwinist paradigm is unsatisfactory in another respect, namely, that it involves only one uniform environment, through which natural selection is exerted in a form which requires specification by one single coefficient for each type of biological entity. Again, as with the omission of the phenotype, there are several different objections to this—though perhaps they should be regarded as different aspects of a basic general objection.

To put the matter abstractly first: there are only two sources of evolutionary change; alterations in the environment, or alterations in genes. A paradigm in terms of a single uniform environment implies either attainment of an equilibrium, or evolutionary changes brought about by the appearance of new genes. But the latter is a very weak prop to rely on, since it is normally held that all possible mutations are constantly occurring at definite frequencies. One could perhaps escape this dilemma by appealing to rare mutational events involving large-scale restructuring of the genotype (additions, deletions, inversions, etc.), or rare incorporation of large masses of genetic information by processes such as introgressive hybridization, incorporation of episomes, etc., but this would be an uncomfortable basis for a general theory of evolution.

On a more pragmatic level, one may ask whether the concept of a single uniform environment is ever even conceivably applicable to the real world, which appears inescapably heterogeneous. And if it is not, it should be remembered that evolution provides mechanisms by which any initial inhomogeneity will become either exaggerated in kind or increased in the number of sub-regimes. For instance, if we start with a total universe containing two environmental regimes (niches) A and B, each dominated by a biological species A' or B', then it will always be possible for some evolutionary descendant of one or other of these species to delimit as its own niche some appropriate function of the previously existing entities, $F(A,B,A',B')$; indeed, there is an infinite set of such functions to be used in this way. This is the general explanation for one of the features of evolution which seems to prove most puzzling to physical scientists, who ask why such an enormous variety of different types should have been produced, although the existence of primitive organisms such as bacteria, at the present day, proves that they are functionally quite 'fit' enough to survive. The point is that their mere presence opens the possibility that something will evolve to exploit them, e.g. as a consumer of them. The utilization, as an environmental niche, of a function of pre-existing species and niches $F(A,B,A',B')$ will not always demand a biological organization more complex than that of A' and B'; a mere parasite on A' may be much less complex or highly evolved than its host. But in general terms one may expect that, among the whole array of functions F which are potentially utilizable, some will actually be used which do require the elaboration of more complex reaction and control systems than any involved in the entities of which the function is composed. We would thus expect to find, not that the evolution of living systems exhibits any universal programme of 'progressive change' (in the sense of increase in complexity or the like), but that there would be a tendency for the highest degree of complexity reached to increase gradually as the process continues. If the primitive biological world was populated only with bacteria, say, it is, for the reasons just given, difficult to imagine it remaining with no more complex organisms appearing within it, even if the combinatorial possibilities of the genes in the bacteria were sufficient to keep a process of evolutionary change on the move at the bacterial level from those times to the present.

BASIC EQUATIONS OF THE NEO-DARWINIAN AND POST-NEO-DARWINIAN PARADIGMS

The paradigm of the fundamental evolutionary process which has been put forward above leads to an algebraic formulation rather different than that of the Neo-Darwinian scheme, and there are differences in the types of question which can be meaningfully posed in the two types of situation. The Neo-Darwinian algebra was most clearly and simply stated by its originator, J. B. S. Haldane[2], and for purposes of contrasting the paradigms we may imitate the extreme simplifications of his exposition.

Clonal Reproduction

The simplest possible evolving system will comprise two vegetatively reproducing haploid clones, with genotypes A and a. In the Neo-Darwinian scheme, these exist in one environment S. The action of natural selection is represented by assigning to one clone, say A, a selection coefficient of 1, while the other clone a has a coefficient of

$1-k$. Then clone A will increase in frequency in the mixed population, and clone a will decrease in frequency.

In the 'post-Neo-Darwinian' scheme, we have to consider not only two clones A and a, but also two environments X and Y, which can affect the phenotypes of members of A and a developing within them. Moreover, we have to allow for the possibility that an organism developing mainly in one environment may be selected mainly in the other. For simplicity, we may at this stage separate these two actions of the environment, and consider the organisms as classifiable into those developed in X and selected in X, those developed in X and selected in Y, and so on. Thus, for each clone there are four classes to which selection coefficients have to be assigned. The behaviour of the system will depend on the attribution of the coefficients to the classes. The kind of questions we can ask of the algebra concern the evolutionary advantage and disadvantages of various principles which might govern the magnitude and distribution of the coefficients.

Suppose there are two clones A and a, and two environments X and Y, with frequencies p and $1-p$. As case 1 let us assume that a proportion q of organisms is selected in X and $1-q$ in Y. Then for each clone we have:

developed in X	developed in X	developed in Y	developed in Y
selected in X	selected in Y	selected in X	selected in Y
frequency pq	$p(1-q)$	$q(1-p)$	$(1-p)(1-q)$

Suppose clone a has perfect adaptiveness, i.e. always has full natural selective efficiency in the environment in which it developed. Then its coefficients would be:

$$1 \qquad\qquad 1-k_1 \qquad\qquad 1-k_2 \qquad\qquad 1$$

But let clone a be fully canalized for X, i.e. show full natural selective efficiency in X whatever environment it had developed in. Its coefficients would be:

$$1 \qquad\qquad 1-k_3 \qquad\qquad 1 \qquad\qquad 1-k_4$$

Simplifying further, we may assume that the chance of being selected within an environment is proportional to the frequency of that environment, i.e. $q=p$. Further, let us take $k_1=k_2=k$ and $k_3=k_4=k'$.

Then if the frequencies of A and a in generation n were $1-u$ and u, respectively, in generation $n+1$ they will be:

$$
\begin{aligned}
A_{n+1} &= (1-u)\,[p^2+(1-p)^2+2p(1-p)(1-k)] \\
&= (1-u)\,[1-2p(1-p)k] \\
a_{n+1} &= u\,[p^2+p(1-p)+(p(1-p)+(1-p)^2)\ k'] \\
&= u\,[1-(1-p)k']
\end{aligned}
$$

Therefore

$$u_{n+1} = \frac{u[1-(1-p)k']}{u[1-(1-p)k']+(1-u)[1-2p(1-p)k]}$$

$\Delta u = u_{n+1} - u_n$ is positive if

$$u[1-(1-p)k']-u^2[1-(1-p)k']-u(1-u)[1-2p(1-p)k]$$

is positive, i.e. if

$$1-(1-p)k' > 1-2p(1-p)k$$
$$k' > 2pk$$

Thus, as might have been expected, which clone is favoured depends not only on the selection coefficients but on the frequency of the environments, and the larger the frequency of environment X, the more likely it will pay to canalize for it.

Mendelian Recessive in a Diploid

This is the classical paradigm case. In the Neo-Darwinist formulation, one assumes a fully recessive gene a in frequency u. Then the array of zygotes in generation n is $(1-u)^2$ AA, $2u(1-u)$ Aa, and u^2 aa. In generation $n-1$ this will be changed to $(1-u)^2$ AA, $2u(1-u)$ Aa, $u^2(1-k)$ aa.

In the post-Neo-Darwinist scheme, we have to envisage two environments X and Y in frequencies p and $1-p$. We can make the same simplifying assumption that both development and selection occur in these environments in proportion to their frequencies. We have to assign selection coefficients to the phenotypes derived from all three genotypes in the different combinations of development and selection. These would be as follows for a case in which the dominant gene produces a fully adaptive development, while the recessive produced canalization for environment X.

	frequency	developed in X selected in X	developed in X selected in Y	developed in Y selected in X	developed in Y selected in Y
AA	$(1-u)^2$	1	$1-k$	$1-k$	1
Aa	$2u(1-u)$	1	$1-k$	$1-k$	1
aa	u^2	1	$1-k'$	1	$1-k'$

It is easy to show that the zygotic frequencies in the next generation will be:

AA $(1-u)^2[1-2pk(1-p)]$
Aa $2u(1-u)[1-2pk(1-p)]$
aa $u^2[1-k'(1-p)]$

Whence

$$u_{n+1}-u_n=\frac{u^2(1-u)\left[(1-k'(1-p))\right]-\left[1-2pk(1-p)\right]}{(1-u)^2\left[1-2pk(1-p)\right]+u^2\left[1-k'(1-p)\right]}$$

This is positive if k' is less than pk.

Alternatively, one may consider the situation in which AA and Aa are canalized for environment X, while aa produces an adaptive phenotype. The selection coefficients will then be:

	developed in X selected in X	developed in X selected in Y	developed in Y selected in X	developed in Y selected in Y
AA and Aa	1	$1-k$	1	$1-k$
aa	1	$1-k'$	$1-k'$	1

From this it turns out that $u_{n+1}-u_n$ is positive if $2pk'$ is less than k. Thus, if environment X is the more frequent one (p greater than 0·5), a gene producing canalization (case 1) can make its way against an adaptive gene in the face of a less favourable ratio of selection coefficients than can an adaptive recessive competing with a canalizing dominant.

CONCLUSIONS

It does not seem appropriate to carry the mathematical analysis any further at this time. It has been offered here, not on the grounds that it is immediately applicable or verifiable, any more than is the classical Neo-Darwinist analysis. What I wished to do was to exhibit a scheme of basic ideas which directs attention towards, rather than away from, the problems which are of most importance for evolutionary theory at the present time. By far the greatest advance in our knowledge of evolution which has occurred in recent years has been the discovery of the enormous range and variety of genetic variation which is present in natural populations. It seems certain that one of the important determinants of this situation is the fact that such populations exist in heterogeneous environments, so that the applied selection criteria are not the same for all individuals. Again, the two major, long-standing problems of evolution are speciation and adaptation. It is generally accepted that a splitting into two taxa involves the operation of two different environments (probably usually allopatric, but possibly in some circumstances sympatric), while an understanding of adaptation demands a theory of how adaptive characters change when the environment changes. In all these three major contexts, the conventional Neo-Darwinist paradigm of the phenotypeless genotype in a uniform environment tends to lead thought away from the challenging questions. [1968]

24

THE BASIC IDEAS OF BIOLOGY

I shall try in these lectures [The Ballard Mathews lectures, University of North Wales, Bangor] to step back a little distance from all the detailed problems presented by living systems, even though they are so fascinating that they can keep a working biologist enthralled throughout a long active life. What I want to do now is to look at biology as a whole and try to discern what are the basic ideas characteristic of this type of science. It is probably best to begin boldly by asking the brash question, 'What is life?' In the fairly recent past most authors have echoed the sentiments of Sherrington about this: 'To ask for a definition of life is to ask a something on which proverbially no satisfactory agreement obtains.' However, this did not entirely prevent them giving their opinions about the matter. And in recent years the question has become very acute again. When we obtain actual specimens of the surface constituents of the moon or Mars or even fairly detailed televised information about them, what sort of evidence would we accept as establishing the existence of anything worthy to be called 'life'?

During at least the first third of this century the most authoritative biologists saw the fundamentals of life in the characteristic ways in which living organisms operate. A living thing, they pointed out, takes in a number of relatively simple physico-chemical entities or molecules from its surroundings (incident light energy, water, inorganic salts, relatively simple organic food materials, etc.) and builds these up into its own structure, in the process carrying out a synthesis of molecular structures much more complex than those it had absorbed. 'The constant synthesis, then, of specific material from simple compounds of a non-specific character is the chief feature by which living matter differs from non-living matter' was the way it was expressed by Jacques Loeb in 1916. The same point of view was put by another leading biologist of that day who represented quite the other end of the spectrum from Loeb in relation to the controversy about vitalism and mechanism, namely J. S. Haldane: 'The active maintenance of normal and specific structure is what we call life, and the perception of it is the perception of life. The existence of life as such is thus the axiom upon which scientific biology depends.'

This view of the basic nature of biology was the inspiration of the conscious movement for the development of a 'modern biology' which took place in Great Britain at the end of the 1920s and early 1930s. This took as its aim the study of 'the nature of living matter', which was the title of a book by one of its leaders, Lancelot Hogben, and it focused on Loeb's idea that there is a special type of substance, living matter. Some authors simply identified living matter with the cell. Sherrington, a few lines after the quotation given above, goes on: 'There is a little aggregate of atoms and molecules such as the world we call lifeless nowhere contains. It is confined to living things; many of these it enters into as a unit, and builds up. It is a unit with individuality . . . To these units Robert Hooke of the Royal Society, an early observer with the microscope, gave the name "cells".' The most influential leader of the modernists, James Gray, states that: 'It seems logical to accept the existence of matter in two states, the animate and the inanimate, as an initial assumption.' However, he did not suppose that the cell is the unit of matter in the animate state. His reasons for this were partly empirical, the fact that one can find rather exceptional organisms in which the cellular organization is not developed (for instance, multinucleate algae or slime moulds) but, more importantly, because the cell is such a comparatively large entity containing so many molecules that it must be subject to statistical regularities and thus under the control of the second Law of Thermodynamics. One of the most profound characteristics of living systems, regarded as working machines, is that they lead to a local increase of order by taking in simple molecules and building them up into complex compounds which are arranged in an orderly fashion. At first sight, this might seem to imply that the system acts against the dictates of the second Law. Gray therefore supposed that 'the real unit of life must be of a protoplasmic nature irrespective of whether it is sub-divided to form a mechanically stable system or not: in other words, cellular structure is not in itself of primary significance . . . It is by no means impossible, however, that the essential units of living matter are each composed of a very large number of molecules and under such circumstances the number of units present may be reduced to a figure which will materially affect the validity of statistical laws.' As these quotations make clear, the refusal to accept the cell as the basic unit of life left intact the notion that life is to be defined as a particular state of matter.

This conception underlay the growth of modern biology conceived of as the experimental study of living organisms as causally operating systems. Within a few decades there was a great increase in our understanding of such processes as muscular contraction, glandular secretion, nervous conduction, the metabolic transformation of molecules into one another, and the whole functional machinery by which a living thing works. It gradually became clear that in nearly all these processes the fundamental role is played by the proteins, which are enormously complex molecules that are found only in living (or recently dead) biological entities. Perhaps the greatest triumph of this type of biology has been to elucidate their structure. It was first shown that they are built up by attaching smaller molecules of amino acids in sequence one to another to form a long string-like structure. Each such string of amino acids is folded together into a complex tangle, and the effective operations of the protein depend essentially on the precise shape that this tangle takes. It is only in the last few years, thanks to the work of X-ray crystallographers such as Astbury, Pauling, Bernal, Perutz, and Kendrew, that we are at last beginning to get precise information about the shapes of these tangles, the forces that hold them together, and the degree to which they can be modified. If life were really no more than a particular state of matter, these studies would be coming very near to giving us an understanding of the basic factors on which the 'nature of living matter' depends.

However, it was not long before the adequacy of the 'living matter' definition of life was seriously challenged. It was pointed out, chiefly by geneticists, that living things do not merely synthesize specific structures out of simpler molecules; it is an equally important fact that they reproduce themselves, and indeed it might be claimed that the most important fact about them is that they take part in the long-term processes of evolution. The capacity to reproduce, that is to give rise to a new unit essentially similar to the old one, demands not merely specific synthesis but the ability to pass this specificity from the initial unit to a new unit which is its offspring. For evolution to be possible something still further is required. It is necessary that changes occur from time to time in the specificity of an organism and that, when such changes occur, they are passed on to the offspring. This will amount to the occurrence of hereditary variation, and Darwin's argument for the inevitableness of natural selection will then be sufficient to ensure that evolution will take place. H.J. Muller, the most farsighted of the early geneticists, argued as far back as 1916 that this provided the only basis for a satisfactory definition of life. A system is living if it carries specificity and can transmit this specificity to offspring and if, in addition, the specificity can change and the changed specificities are also transmitted.

Nowadays, using modern jargon, we should rephrase this slightly, using the word information instead of specificity: a system is living if it encodes hereditarily transmittable information, if this information sometimes suffers alterations, and if the altered information is then transmitted.

The geneticist's definition of life is obviously much more general than the physiologist's definition given above. It does not connect life essentially with the particular types of living matter which we can inspect on the earth's surface. In point of fact on the earth's surface hereditary information is, so far as we know, always encoded in a class of compounds known as the nucleic acids and, in our experience, this is always expressed by the formation of corresponding proteins which carry out the functions which the physiological biologists have investigated and described. But the genetical definition does not require that this should be so. We might have genetic information encoded in other compounds and worked out in a different way. This opens a much wider range of possibilities when we contemplate the possibility that there exists something worthy to be spoken of as 'life' on other planets or stars in which the physico-chemical conditions are totally different from those on the surface of the earth. Moreover, it makes it possible to discuss, in a meaningful way, the problem of the origin of life from inorganic materials. For these reasons, it has, in recent years, tended to be more and more widely adopted, and the old idea that there is a characteristic type of 'living matter' has passed more and more into the background.

In my opinion, this reaction, although basically justified, is in danger of going too far. I do not believe that if we discovered systems that were nothing more than mechanisms for the hereditary transmission of mutable information we should, in fact, consider them to be living. Such systems would show none of the properties included in the older physiological outlook, which regarded the characteristic of life as specific synthesis and an apparent local reversal of the second Law of Thermodynamics. One can get some idea of their character by considering the well-known game in which you are given a word, say Bit, and are required to change this into something else, say Man, by altering one letter at a time in such a way that the new combination always spells an accepted English word; and we could, if we liked, expand the rules so that letters could be duplicated or pieces of the word reversed in order or transposed from one place to another, etc. We have perhaps a system rather like this in the growth of crystals. In a normal perfectly-formed crystal new molecules are laid down in regular order on the face of a pre-existing arrangement of molecules, but if a mistake happens and one new molecule gets into a slightly wrong position this frequently produces an irregularity on the surface which affects the probability that a new molecule will settle down at that place. Thus the mistake is, in some sense, transmitted in that it affects the next generation of molecules. If the irregularity is such as to increase the chance of a new molecule being deposited there, one could regard this as a selectively favourable effect. We can thus have processes essentially akin to the natural selection of mutable hereditarily transmitted information, but we do not regard such systems as alive, and I think the reason is essentially because they are not interesting enough! They fulfil Muller's criterion but they do absolutely nothing else. To be worthy of being called alive they must, I think, exhibit some sort of 'physiological activity'. This need not of course be of the kind exhibited by living things on this earth. They might produce 'living matter' of a totally different chemical composition and mode of operation from that which we find around us, but I should argue they need to do something more than mere information-transmission.

Muller defined life entirely in terms of what geneticists call the genotype. The argument I am putting forward amounts to saying that life involves not only the genotype but also the production of something of the kind geneticists speak of as the phenotype, that is to say, something which is developed out of the genotype and which interacts with the surrounding non-living environment. It was the phenotypes produced on earth that the physiologists were considering when they defined life in terms of a state of matter. Although one cannot accept this geocentric limitation, there is, I think, a good case for insisting that a living system must have some sort of phenotype or other.

THE THEORY OF PHENOTYPES

Whether or not we accept the argument that life necessarily involves not only genotypes but also phenotypes, there is no doubt that all the living systems we know on this earth do have a phenotype as well as a genotype—even viruses. The theory of phenotypes is therefore an essential part of the general theory of biology. It will be discussed under a number of separate headings.

Information theory. Around 1948 Shannon and Weaver developed an elegant mathematical theory for handling certain problems in which the point at issue was the amount of variety or specificity contained within a system. There are very many biological problems where we might be tempted to use this same form of words—amount of

variety or specificity. For instance, we have seen that the physiological definition of life was phrased in terms of 'specific synthesis'.
Again, we speak of genetic variation, or say that the genotype
specifies the nature of the phenotype. There has therefore been a
great temptation to use the Shannon-Weaver theory in connection
with biological problems. However, their 'information theory' was
developed in connection with a particular type of process and has
limitations which make it extremely difficult, if not impossible, to use
it in many of the biological contexts to which people have been
tempted to apply it.

The theory was developed (in the Bell Telephone laboratories)
in connection with the problem of transmitting a message from a
source A through a channel B to a receiver C, and the basic purpose
for which it was evolved was the question of how the characteristics
of the channel B influenced the amount of information that can be
transmitted in a given time. One of its basic results is that, in a *closed
system* into which nothing comes from outside, you can never get
more information into C that was originally contained in A, although
you can, of course, change the form of information, say, from the
dots and dashes of the Morse code into letters written with a typewriter. Now there are several biological situations which present a
close parallel to the systems investigated by Shannon and Weaver.
The most obvious example is the passage of electrical impulses
through networks of nerves, where one has an almost precise parallel
to the transmission of electrical impulses along telephone wires. Here
their theory and its elaborations have proved extremely valuable.
A slightly more far-fetched analogy can be found in the transmission
of the hereditary information contained in the chromosomes of one
organism to the chromosomes of its offspring. But here already we
find that biology has developed mechanisms more flexible than those
used by telephone engineers. There is a system by which the transmitted information can be changed—a gene may mutate so that
what the offspring receives is not exactly the same as was contained
in its parent. Again, there are mechanisms of chromosomal deficiencies, duplications, translocations, formation of isochromosomes,
etc., by which the amount of information can be either increased or
decreased. However, the information theory language is useful
insofar as it allows these situations to be clearly expressed.

It is when we come to consider the relations between the genotype
and the phenotype that the limitations of the theory become of
overriding importance and rapidly render it not merely useless but
a dangerous snare. In the very first steps of the transition from

genotype to phenotype it can still be applied. The genes, composed of DNA, begin to make their effects felt by serving as the patterns (templates) on which there are synthesized the messenger RNAs. In these there is, for every sugar-nucleotide unit in the DNA, a precisely corresponding sugar-nucleotide unit in the RNA. The change is no more than, as it were, changing from a Roman face type to an Italic type, using the same alphabet. It is usually spoken of in biology as a 'transcription'. Even in the next step there is nothing very much to be alarmed at. The sequence of sugar-nucleotide units in messenger RNA provides the pattern for the laying down of a corresponding sequence of amino acids into a polypeptide chain that forms a protein. This involves a more drastic alteration, perhaps comparable to changing from a normal alphabet to one in Morse code. It is usually spoken of in biology as a 'translation'. It is meaningful to discuss the whole of this sequence in terms of information theory and to ask whether any information is lost in the processes of transcription or translation, and to pose questions about the 'code' according to which the RNA nucleotides are translated into the protein amino acids.

It is in the next stages of the formation of the phenotype that information theory becomes unable to deal with the situation. It is obvious that the phenotype of an organism does not consist simply in a collection of all the proteins corresponding to all the genes of the genotype and nothing else. It is instead made up of a highly heterogeneous assemblage of parts, in each of which there are some, but not all, of the proteins for which the genes could serve as patterns, and in each of which there are also many other substances and structures over and above the primary proteins corresponding to particular genes. Each organ—liver, kidney, brain, etc.—is in the first place like a Bridge hand, containing only a certain sample out of the whole pack of cards, and further, contains a whole lot of things—lipids, carbohydrates, pigments, and what have you—that are secondary accretions which one cannot consider as having been in the original hand when it was first dealt. Information theorists have struggled in vain with this problem. It seems quite obvious to common sense that a rabbit running round in a field contains a much greater 'amount of variety' than a newly fertilized rabbit's egg. How are we to deal with the situation in terms of an 'information theory' whose basic tenet is that information cannot be gained? There have been several attempts:

(a) Raven has argued that we have underestimated the amount of information contained in the egg because we have considered only the information contained in the chromosomes, i.e. the genes. He argues that there is also information contained in the cortex of the egg. Taking the area of this cortex and an estimate of the size of a biological molecule, he comes up with another quantity of information which can be multiplied into that in the chromosomes, and he tries to show that this is not too far away from the amount of information contained in the adult organism, as estimated by taking its size in terms of the size of the molecules of which it is composed. But the whole set of estimates here is so fantastically imprecise—Raven considered the cortex of a small egg but could have got himself another factor of 10^4 or 10^5 if he had taken an egg of the size of a hen's ovum—that the whole procedure becomes totally unconvincing.

(b) One can also point out that the system we are dealing with is not a closed one. An organism, as it develops from a fertilized zygote to an adult phenotype, nourishes itself by taking in food from its surroundings. Now the molecules of food—proteins, amino acids, fats, carbohydrates, etc.—all contain some chemical information over and above the bare atoms of which it is composed. Could the adult rabbit have got its extra information, additional to that in the zygote, from the grass it ate? I frankly don't know the answer, but it sounds to me like suggesting that you could add to the information in a paragraph of prose by using a more curly serifed typeface instead of simple sanserif designed on the basis of nothing but straight lines and segments of circles.

(c) But the basic point is surely that in the transition from the zygote to the adult the 'information' is not merely being transcribed and translated but is operating as instructions—if you want to put it in fancy jargon, as 'algorithms'. The DNA makes RNA and the RNA then makes a protein and the protein then does something to its surroundings, which results in the production of more varieties of molecule than there were before. There is nothing very mysterious in this, unless you try to see it in terms of messages going down telephone wires. Suppose you put together in a reaction vessel two chemical species A and B, where A is a molecule having NH_2 groups at a few sites—say, 1, 3, 9—and B is a molecule with the ability to substitute an OH group for an NH_2 group. After a time there will be a much greater variety of molecules in the reaction vessel than there were to start with. The amount of 'information' will have increased, but there is no reason why we should feel that we are witnessing a phenomenon which contravenes the fundamental laws of the physical universe. All we need to realize is that we are dealing with a situation outside the very narrow limits to which information theory applies.

This is, so far as I can see, all that is meant by Elsasser's argument that a comparison of the information content of a fertilized egg and adult phenotype makes it necessary to postulate the existence of 'biotonic laws'—what this boils down to is that information theory cannot usefully be used in considering the relation between genotypes and phenotypes. As I put it some years ago, a genotype is like a set of axioms, for instance Euclid's, and a phenotype like a three-volume treatise on Euclidean geometry, which proves Pythagoras's theorem in the liver, and in the kidney that a tangent to a circle is perpendicular to the radius at the point of contact, in the lungs that the three angles of a triangle add up to 180°, and so on. The inapplicability of information theory to such situations may be demonstrated by pointing out, firstly, that a theorem is proved by showing that it is a tautological expansion of the axioms and therefore contains nothing that was not in the axioms to begin with, but that, secondly, giving such a demonstration is by no means without significance, since if you could prove Pascal's theorem that $a^n + b^n = c^n$ is impossible for any value of n greater than 2 you would be assured of undying fame among the tribe of mathematicians.

In sum, information theory in the strict sense is not a useful tool in considering the relation between the genotype and the phenotype, although the word 'information' is, if used fairly loosely, quite a useful expression to employ in place of such phrases as 'amount of variety' or 'specificity'.

The phenotype as an epigenetic phenomenon. At the beginning of all textbooks of genetics, a very necessary distinction is drawn between the genetic potentialities which an organism inherits from its parents and the manner in which these potentialities become actually realized. The former is the genotype, the latter the phenotype. It is common, even in the best and most recent textbooks, to describe the phenotype in an extremely incomplete and over-simple manner. For instance, Srb, Owen, and Edgar are content with saying: '*Phenotype*, a word that refers to the appearance of an individual'. Even such a cautious and farsighted author as Stern, after pointing out that: 'A character or trait may be defined as any observable feature of the developing or fully developed individual', goes on to say: 'For the genetic constitution, the term *genotype* has been coined; for the external appearance, the term *phenotype*.' Mayr comes nearer to getting the matter into proper perspective when he writes: 'Our ideas on the relation between gene and character have been thoroughly revised and the phenotype is more and more considered, not as a mosaic of individual gene-controlled characters, but as the joint product of a complex interacting system, the total epigenotype (Waddington, 1957).' Even Mayr, however, does not clearly express the most fundamental and basic characteristic of phenotypes, namely that they change in time. The phenotype is the name given to the results of the activities of genes. In a very simple organism, these activities may be carried out in a relatively short period, resulting in the formation, for instance, of a certain number of proteins. In more complex organisms, these proteins themselves interact with one another and with other substances so that it is a long and complex sequence of processes that the genes set in motion. But in either case a time duration, whether short or long, is an essential component of a phenotype.

One can use the expression 'a phenotypic character' (or character for short) to refer to any particular aspect of a phenotype which is being singled out for attention. Some years ago (1947) I introduced the word 'epigenetics', derived from the Aristotelian word 'epigenesis', which had more or less passed into disuse, as a suitable name for the branch of biology which studies the causal interactions between genes and their products which bring the phenotype into being. The word is now quite often used in that sense, but unfortunately it seems to be an attractive expression, and some other authors have seized on it to refer to quite different conceptions. For instance, Elsasser wants to use the adjective 'epigenetic' to indicate that in a certain process information is not being conserved according to the orthodox Shannon-Weaver theory. In my opinion it would lead to greater clarity if the word is kept for the causal study of development, the meaning originally suggested for it.

If one wishes to formulate the phenotype in mathematical terms it is clear, then, that it is a function which involves the time variable. Moreover, the function must involve something more than merely the three dimensions of space, since we are interested in something more than the bare geometry of the organism. We shall need in fact one variable for each constituent (chemical or geometrical) of the system which is relevant to the questions being considered. We shall therefore usually be faced with a function containing very many variables. In order to represent this geometrically we should have to resort to a multi-dimensional phase space to accommodate all the variables of constitution, and within this space the phenotype will be represented by some kind of figure, which will begin at the point representing the constitution of the egg and will then be extended along the time dimension. Theoretically this figure might take the form of a bounded continuous sheet, for instance that of a triangle. If this were the case, one should find that, at some time after fertilization, the phenotype exhibited continuous variation in composition as one passed from one position within it to another. It is an empirical observation that this is normally not the case. In the organisms that we come across, we usually find a number of discrete and distinct organs—a liver clearly marked off both spatially and in composition from the kidney, and both of these from the heart, and so on. This means that the figure representing the phenotype has to branch out into a number of separate sub-configurations, each of which extends separately forward along the time dimension. Little generality is lost if one represents each of these sub-configurations by a line. One can therefore represent the phenotype as a branching system of time-extended trajectories in the phase space.

Although these points may seem obvious, and were stated as long ago as 1940, it is doubtful how far they yet form a part of the working system of thought of most biologists. For instance, one still often sees the remark that cell differentiation during development is explained by the switching of cells into 'alternative steady states', the idea being often attributed to Delbrück. But if one considers any actual case, such as the switching of part of the ectoderm of an amphibian gastrula to become neural tissue or to become epidermis, it is quite clear that it does not enter a steady state but, on the contrary, sets off on some particular pathway of change which, for instance, in one direction will lead it to form part of the neural plate, then the neural tube, then some region of the spinal column, or brain, forming part of the grey matter or white matter or perhaps growing out as a nerve trunk. It is only after it has passed through a large number of transient stages that it gets into a situation which is relatively stable for a considerable period of time, namely the adulthood of the individual.

The elementary processes of epigenetics. Epigenetics has two main aspects: changes in cellular composition (cellular differentiation, or histogenesis) and changes in geometrical form (morphogenesis). How far can we specify what are the elementary processes to which the more complex examples of these phenomena should be reduced?

1. About morphogenesis we can, I believe, say very little. Right at the bottom of the 'ladder of causes' there must be such processes as: (a) the determination of tertiary structures of proteins by the nature of their primary amino acid sequence, and (b) the association, often by still weaker types of bonding, of macromolecules (possibly

sometimes with ordinary molecules as mortar between the joints)
such as we see it in the reformation of striated collagen or myosin
fibrils from solution, the formation of myelin figures or other sheet-
like structures, in systems involving immiscible phases, etc. But there
is an enormous gap between such phenomena and, for example, the
appearance of just five digits of particular lengths and shapes in a
developing limb. I do not think we can indicate any one type of
elementary process (or even a small number of them) which provide
a general pattern for the intermediate steps. It seems rather likely
indeed that these steps are carried out by a large number of different
types of mechanism and that there is no one unitary theory of the
way morphogenesis is brought about.

2. We can probably go somewhat further in relation to cellular
differentiation. It is in fact conventional to say that the basic
elementary process is the derepression (or possibly switching on)
of a structural gene by means of a cytoplasmic gene-recognizing
('genotropic') substance which has been produced by some other
genetic locus. The notion of the derepression of single genes has been
derived primarily from work on bacteria, and the question arises
whether or not there is another more complicated level of elementary
process which is basic for the cells of higher organisms, in which the
DNA of the chromosomes is normally combined with protein. As far
as I know, there is no evidence that you can ever switch on single
genes in higher organism cells, except in the terminal phases of
development when a lot of other correlated genes have already been
switched on (e.g. you can switch on and off the haemoglobin gene in
cells of the erythropoetic series but not in kidney cells or nerve cells).

I have argued that the elementary differentiation process of a
higher organism cell (a) involves complexes or 'batteries' of genes
rather than single genes; (b) takes place in three phases: acquisition
of competence, in which several different batteries of genes become
ready to enter the next phase; determination, in which one of these
batteries is singled out to become dominant in the future history of
the cell; activation, in which the proteins corresponding to the
structural genes in this battery actually begin to be produced. Since
cells can divide while remaining in phase 2 (determined but not
activated), it seems almost certain that determination must be a
process operating at the gene level. It is not clear, however, whether
it leads to the production of corresponding messenger RNAs, or
whether this is always the first phase of activation; even if the latter
is the case it must be admitted that complete activation may also
involve control of protein synthesis at the ribosome level.

The canalized or buffered character of the epigenetic trajectories. It is an empirical observation—but one of profound importance for much of biological theory—that epigenetic trajectories normally show some resistance to being changed. The evidence is of two main kinds: (a) Developmental: It is extremely common to find that a developing system is, at least for some periods of time, capable of 'regulation', in the sense that it is capable of 'compensating for' disturbing influences and returning to normality at later stages in the developmental process. (b) Genetic: It is again very usual to find that slight changes in a genotype may produce no deviation in the developing phenotype (e.g. in the phenomena of dominance, epistasis, etc.).

Phenomena involving the holding constant of some parameters of a physiological situation (e.g. the oxygen tension or pH of the blood) have been well known for a long time. The situation is usually referred to as one of 'homeostasis'. We are here dealing with a similar concept, but of a rather more general nature, in that the thing that is being held constant is not a single parameter but is a time-extended course of change, that is to say, a trajectory. The situation can therefore be referred to as one of homeorhesis, i.e. stabilized flow rather than stabilized state.

(I should like to see some mathematician express this contrast in more precise terms; for instance, homeostasis does not in general involve the system returning to a single point in phase space. In order to keep constant the oxygen tension of the blood you may increase the rate of heart-beat. I suspect that the situation is something like this: in homeostasis the phase space contains what I think Thom would call an 'attractor line', in the equation for which one variable had a constant value; in homeorhesis the attractor line is parallel to the time axis and none of the variables are necessarily constant.)

The name 'chreod' has been suggested to refer to a canalized trajectory which acts as an attractor for nearby trajectories. It is an interesting question to discuss how far the existence of chreods is necessary and how far it is merely an empirical result of the operation of natural selection. There are several points which I do not quite know how to put together:

1. In many of the chreods we meet in actual animals, both the value of the parameters and the degree of stabilization is under the control of natural selection. For instance, in selection experiments involving body size in *Drosophila*, Forbes Robertson found good evidence for the buffering of final body size to particular values, but the actual

magnitude of these values was different in different selection lines. Again, it is a common observation that the introduction into a relatively normal genotype of some strongly acting mutant (which does not persist in natural populations so that its effects can be subject to natural selection) often produces a phenotype which is not only highly abnormal but also highly variable, i.e. it destabilizes the chreod. Finally, artificial selection can certainly build up new chreods, as in the experiments on genetic assimilation.

2. However, this does not seem to me to rule out the possibility that the existence of some sort of chreod or another is a necessity. If one mixes together a large number of active agents (say chemical substances) which can act upon each other, the mixture will change through some defined course in time. The question is, will this course show any properties of buffering or canalization? I should have guessed intuitively that provided a large number of the components can interact not only pairwise but each with a number of other components, some degree of buffering would be bound to emerge [1].

3. When any of the interactions between the components are strongly non-linear—crudely speaking, involve threshold phenomena— then some sort of chreodic behaviour seems to be inevitable [see further discussion of this in paper 22, p.197].

Another set of questions concerns the type of mechanism by which the buffering is produced. It is becoming conventional to refer in this connection to 'feedback'. Strictly speaking, as I understand it, feedback refers to a situation in which, if process A deviates from normal, the system reacts in such a way as to bring process A back to normality. In biology something of this kind does occur in such processes as end-product repression and end-product inhibition, in which the final product of an enzymatic pathway acts so as to control various earlier steps in the sequence by which it is synthesized. However, there is another and perhaps more usual type of buffering action in which, when process A deviates from normality, nothing happens to bring it back again, but the system as a whole simply absorbs its effects. For instance, Kacser has shown that if a mutation drastically reduces the efficiency of one enzyme in a sequence of synthetic steps, the consequence may merely be an increase in the concentration of its immediate precursor, leading to the same overall flow rate through that synthetic step and complete absence of any effect on the later steps. Much more complex compensatory reactions could, of course, be envisaged in networks, rather than simple sequences, of chemical reactions. This is absorptive buffering rather than feedback in the strict sense.

There is, however, no doubt that feedback dependent on direct coupling between processes does often occur. In particular, the regulator-operon system of biochemical linkage between genes is likely to give rise to many 'strong' interactions between cellular processes. The general theoretical consequences of these have been investigated by Goodwin. I shall not attempt to summarize, except to point out that his argument, that oscillatory phenomena are to be considered as the normal pattern of cellular biochemistry, opens a great many new avenues of thought in fundamental general biology. Nobody in the past has, so far as I know, conceived of a cell as being *essentially* an oscillator, and the idea opens a whole new range of possibilities concerning the interactions between cells. At this time I should like to make only two remarks: (a) As I understand it, the algebraic analysis is only explicitly soluble if one makes the simplifying assumption that there are only binary interactions between constituents. This seems to amount to the assumption of what I called above 'feedback in the strict sense'. Does the conclusion about oscillatory behaviour still hold if the canalization of a chreod depends on absorptive buffering which involves interaction between very many components? (b) It is not obvious that the new system of ideas, important though they may be in theoretical biology, have any major implications for general philosophical questions.

The importance of understanding the nature of the mechanisms on which chreodic behaviour depends arises from the fact that a chreod, as we have defined it, is simply the most general description of the kind of biological process which has been referred to as 'goal directed'. The nature of such processes has always been recognized as one of the major problems of theoretical biology. The words to be used for describing them and discussing them are still matters for debate. The earlier expressions 'teleological' and 'finalistic' are usually thought to carry an implication that the end state of the chreod has been fixed by some external agency and that the end state is in some way operative in steering the trajectory towards itself. To avoid such implications I have spoken of such phenomena as 'quasi-finalistic'. and the word 'teleonomic' (introduced I believe by Pittendrigh in *Behaviour and Evolution*, 1958) has been used as a substitute for teleological. On the whole, however, I believe it is preferable to use words (such as chreod) which do not lay such stress on the final state but draw attention to the whole time-trajectory.

The two main discussions of the topic in recent years by professional philosophers of science are, I think, those by Nagel and

Sommerhof. These are both set out in fairly technical language, but when one penetrates this one finds, first, that both of them have discussed situations of a quasi-finalistic kind, i.e. concentrating on the attainment of an end state rather than on following a trajectory; and secondly, as far as I can see, the crucial problem is merely smuggled-in as one of the assumptions underlying the logical algebra without being actually explained in any way. For instance, Nagel assumes the existence of two sets of 'state variables' B and C which are so related that variations in one are balanced out by corresponding variations in the other. Sommerhof indulges in essentially the same procedure with his two variables R and E. To my mind these amount to no more than statements of the problem, not explanations of them; statements moreover in a form which combines the disadvantages of being so abstract as to give no hint of a direction in which an explanation might be sought, and of being of such narrow scope that it omits the fact that we are dealing with trajectories not merely with end states. I think therefore that the plain language discussion given above penetrates further than any of the existing 'metamathematical' discussions, though it will I hope soon be superseded by some more sophisticated treatment (possibly derived from topology?).

Great hopes are often held out for the application to biological problems of new mathematical approaches grouped under the general heading of 'Systems Theory'. Looking at articles such as, for instance, that by Quastler, the sceptical biologist is likely to ask, firstly, what is the real gain in either flexibility or precision in substituting the terminology input – black box – output for the more old-fashioned stimulus – cell – response? And secondly, what actual theories in any way applicable to biology have emerged from any of the four major types of systems theory he refers to: cybernetics, games theory, decision theory, and communications theory? In too many of the articles written for biologists on such subjects (including Quastler's) the exposition stops after problems have been formulated in a new fashion and before any theorems have been proved (apart perhaps from some points about the minimax strategy in games theory which Lewontin has applied to some problems of evolution theory).

The nature of biologically transmissible information. The phenotype can be regarded as, in the main, an exposition into a number of different propositions of the information transmitted in the DNA of the chromosomes. We know of some, probably minor, variants of this type of information: (a) Pro-virus-like bodies also consisting of

DNA, which may exist either attached to the chromosome or free from it. (b) RNA viruses in which RNA shows the same capacity for self-replication, i.e. for transmitting information, as DNA does. Must we conclude then that all biological transmission of information is carried out by nucleic acids?

The first question is, have we good evidence for the biological transmission of information by systems lying outside the chromosome? The answer seems to be that we have. Some of this is concerned with the very highly specialized cortex of the ciliates (e.g. Sonneborn on Paramecium) or other specialized cell organelles such as chloroplasts. In both these instances there is a good deal of evidence that nucleic acids are present in the structures, and it *might be* that it is nucleic acid which transmits the information. There is other evidence of non-chromosomal heredity, particularly that of Sager, in which the location of the transmitting system is at present unknown. Again it *might be* that the transmitting agent is a nucleic acid. Indeed, the hereditary system shares some of the detailed properties of chromosomal heredity (e.g. intra-allelic recombination) which makes this seem probable. However, we also know of examples in which cell structures, thought not to contain any nucleic acid, carry 'information', in the sense of specificity which can have an active effect on processes going on in the neighbourhood. For instance, the arrangement of enzymes on the mitochondrial membrane is of this character and, on a larger scale, there is a great deal of evidence for similar operative information in the cortex of egg cells; again one can refer to the growth of cellular organelles, such as the nuclear envelope and stacks of annulate lamellae, where the appearances strongly suggest, though they do not prove, that the existing structural arrangement plays a part in the production of new similar structures in its neighbourhood. There seems to be no good theoretical reason why such information-containing structures should not exist. The main question to be asked about them relates to their capacity for self-replication, that is to say, for information transmission. A very few biologists (e.g. Curtis) think that the cortical information in an egg cell can be transmitted over many generations. The majority of biologists doubt this, but there seems no good reason why such information should not be transmitted through the formation of at least a certain number of replicates as, for instance, in the growth of a mitochondrion or nuclear envelope.

It may be profitable to ask what are the essential requisites for an efficient mechanism of biological information transmission. One most important one would be to make the system independent of the disturbance caused by cell division. DNA has got over this hurdle, since the whole cell division process is obviously carefully organized around the process of DNA replication and the separation of the replicates. Small information carriers, such as small portions of the cortical surface or small organelles within the cytoplasm, could also avoid being disturbed by cell division. If they can operate so as to guide synthetic processes in their neighbourhood so that they build up replicates of themselves, there is no reason why they should not show an indefinite genetic continuity. There may well be cell organelles which have this property (centrosomes, basal bodies of flagellae, etc.?), which, that is to say, have a definite structural organization and cause the appearance of a replica of themselves in their surroundings. However, they would not be information-carrying systems in the full sense unless they could suffer alterations (mutations) and transmit the mutated state. This obviously demands something more than mere capacity for self-replication of a constant structure. It requires, in the first place, that the mutated structure can have a specific effect on the building up of the new replica, that is to say, if the normal structure is ABC, but owing to a mutation a structure ABD appears, it is necessary that ABD should cause another group of A, B, and D sub-units to come together in the proper arrangement. One can envisage a situation in which it might be able to do so provided sub-units D were available, but if the sub-units are themselves produced under the control of some other information system (e.g. chromosomal DNA) the inherent capacity of the ABC to transmit information could not actually be utilized. It seems rather likely that it is for this sort of reason that the DNA system has got such thorough control. It seems possible that cells may be full of entities which could transmit information but only by manipulating sub-units of a complex kind with which the genes do not always provide them. They would be, as it were, capable of manipulating words, but not of building the words they required out of individual letters.

This topic is closely related to what I have called 'template production of non-copies' as seen, for instance, in experimental embryology. In these instances a region of tissue carries a spatial pattern of information which is operative in calling forth the appearance of organs which correspond to, but are not identical with, those from which the information has proceeded. In such systems, however, no question of genetic continuity arises and we will not discuss them further in this place.

Questions about evolution. There is universal agreement nowadays that the foundations of the theory of evolution are to be found in Darwin's concepts of random variation and the survival of the fittest. However, it is as well to notice that modern orthodox neo-Darwinism, although using the same phrases, has actually changed the meaning of almost all the words, so that what emerges is quite considerably different from what Darwin put forward.

1. 'Random variation'. Darwin was thinking of random phenotypic variation. Neo-Darwinism is thinking of random genotypic variation. Here neo-Darwinism is right, in so far that all the evidence suggests that there are no constraints limiting the freedom of mutation (it's just worth considering whether certain nucleotide sequences might be excluded for some reason, but there is little evidence of it). However, neo-Darwinism seems to me to be wrong in so far as it usually tacitly assumes that randomness of genetic mutation implies randomness of phenotypic variation, and I believe Darwin himself was wrong in so far as he believed that phenotypic variation can be adequately characterized as random—it is doubtful if he actually did believe this.

2. 'Survival of the fittest'. Darwin often argued as though he was really thinking about survival in the sense of an organism living for a long period, and he used the word fittest to mean 'most able to carry out the ordinary transactions of life', such as running, collecting food, etc. The neo-Darwinist meaning is quite different. For survival they substitute—quite rightly—reproduction; and by fittest they mean 'most effective in contributing gametes to the next generation'. Thus the whole consideration of ability to carry out the ordinary business of life has disappeared from the neo-Darwinist theory, and is replaced entirely by the conception of reproductive efficiency. This in effect reduces Darwinism to a tautology, and leaves for quite separate discussion—which is very rarely provided—why animals should have evolved all sorts of highly adaptive structures to do unlikely things instead of simply being reduced to bags of eggs and sperm like certain parasitic worms. These comments apply to the formal mathematical theory of neo-Darwinism. The most valuable modern evolutionary thought is not to be found in this form of theory but in more generally phrased discussion, and it is the systematization of these views that offers the most profitable subject for consideration.

One may begin by asking what is the major problem on which the theory of evolution attempts to throw light. Again, the answer is quite different nowadays from what it was for Darwin. For him the major problem was to establish the point that species have changed and are derived one from another. Nowadays everyone accepts this, and there is no point in pursuing it. For us the major problem is one which was only a second order issue to Darwin. This is the problem of adaptation. Why do we find animals and plants which have structures and capacities that make them admirably suited to carry out extraordinary living routines in the most unlikely situations, often highly unfavourable for reproduction? A second major problem for evolutionary theory today is to understand how and why living organisms have become divided up into separate taxonomic categories—and although Darwin called his book *The Origin of Species* he said remarkably little about this point.

Modern biological theory holds firmly to the conclusion that the major agencies producing evolutionary change are random mutation and differential reproduction, with minor complications due to migration and hybridization (more important in plants) and essentially no contribution from environmentally directed hereditary variation. The major new information we have accumulated has been concerned with the genetic structure of the populations of organisms in which these processes occur. We know that these populations contain much more genetic variation than would be apparent at first sight. We know that segregation and recombination is an extremely important source of variation in the short term, though ultimately dependent on random mutation in the background. We know further that the variation in a population is 'co-adapted'. However, these points, although of the greatest importance in setting the stage on which evolutionary theories must operate, do not in themselves directly contribute to the solution of the major problems mentioned above. Moreover, as we have seen in the first paragraphs of this section, the mere invocation of reproductive efficiencies and Malthusian parameters leaves the essential problems on one side.

The remaining avenue left to be explored is that opened up by the remark that randomness of genetic variation does not imply randomness of phenotypic variation. We have, in my opinion, to return to Darwin's concern with the nature of phenotypic variation —a subject about which he was always complaining that there was, in his day, a total absence of any understanding. Nowadays not only has the development of genetics given us some insight into genetic variation, but the development of epigenetics is giving us at least a few hints as to the nature of phenotypic variation.

Some problems for the immediate future of evolution theory

1. The old problem of the relation between heredity and environment in evolution (what is often, though somewhat incorrectly, spoken of as the problem of Lamarckism) has, I think, largely been cleared up by the recognition that the capacity of an organism to respond to environmental stresses during development is itself a hereditary quality. Further, the demonstration that a combination of this fact with that of developmental canalization leads to the occurrence of a process of genetic assimilation, by which the effect of an 'inheritance of acquired characteristics' can be exactly mimicked, has removed the whole heat out of this ancient discussion. These developments of biological theory have, however, I believe, considerable philosophical implications, since they show how it is possible for characteristics such as mental or perceptive abilities, which originally arose in interaction with the environment, to become in later generations 'inborn' to the extent of being independent of any particular environmental stimuli.

2. There has not yet been sufficient exploration—intellectually, let alone experimentally—of the feedback situation, in which an animal's behaviour largely determines the kind of selection pressure to which it will be subjected. Consideration of this will probably need an elaboration of evolutionary theories involving selection between small, semi-isolated populations or 'demes'. In so far as this process occurs, one could consider a large population of animals as playing a game against nature, making a series of moves in each of which one deme would adopt one type of behaviour and submit itself to the resulting natural selective pressures, while other demes might choose to operate in a different way. It is in this connection that games theory might have much to offer evolution theory, if indeed it has any theorems to offer anybody; but so far I have not been able to discover what they are.

3. What points of principle, if any, emerge from the fact that we can now begin to draw up evolutionary trees of the changes in amino acid sequences in proteins? So far as I can tell up to date, studies of this kind have no more (and possibly even considerably less) to tell us about the general theory of evolution than do studies of any other phenotypic characters, though they may have something to tell us about the way in which particular proteins carry out their biochemical operations.

4. Recent evolutionary theory has concentrated very largely on questions of the responsiveness of populations to selection pressures by alterations in gene pool. There has been much less development

of theories concerned with changes in numbers of a population in competition with other ecological competitors, and in relation to a random or even, as Lewontin calls it, a capricious environment.

One of the more immediate tasks of evolutionary theory is to decide what, if anything, could be meant by 'the fitness of a species (or a population)'. In current technical practice, 'fitness' is a parameter ascribed to individuals of a given genotype to generation N, and measures the relative probability that they will produce offspring which succeed in transmitting that genotype to generation $N+1$. That is to say, its definition pays attention only to the very short term, and the concept is clearly inadequate in relation to long-term evolution. Thoday has suggested a notion of 'fitness' which is defined in terms of the chances of leaving offspring 10^{6+} generations hence. This suffers from the opposite drawback, that it has no relevance to evolution within geologically reasonable periods, e.g. none of the competing families of Jurassic ammonites survived the Cretaceous. The real question would seem to be: how can we compare, here and now, the effectiveness with which two populations will probably be able to cope with a future which is essentially not completely foreseeable? The most useful concept, in my opinion, would be one in which 'population fitness' would be defined in terms similar to those which one would use to determine the value of a hand at cards, a set-up on a chess board, or, even better, the 'usefulness' of a tool (an adjustable wrench in comparison with a box spanner, say). But how, if at all, can one do this? Perhaps the inventors of games-playing automata have something to offer?

5. One could, in this connection, also ask the same question as was asked in relation to development: What is the basic elementary process of evolution? According to neo-Darwinism it is simply to leave more offspring than your neighbour. But, as pointed out above, this formulation omits any reference to alterations in any other aspects of phenotype than those concerned with reproduction. A more adequate answer would be something like: To find some phenotypic modification which facilitates leaving more offspring than your neighbour; or, to put it more generally, to find some way of coping with the situation (I realize that Americans say Britons *cope* with situations while they *change* them).   [1968]

25

This essay is dedicated, with respect and affection, to my good friend Theodosius Dobzhansky, in celebration of his seventieth birthday.

The theory of evolution has passed through two main phases. In the first ('Darwinian'), the essence of the process was held to be natural selection operating on a hereditary system characterized by 'blending' inheritance, in which new hereditary variation was brought into being by some unknown mechanism. In the second ('Post-Darwinian'), the hereditary system on which natural selection acts had the Mendelian properties of dependence on discrete alternative states of the hereditary factors and the production of new variation by mutation. The enormous advances made in our understanding of evolution by calling on the resources of Mendelian genetics need no emphasis.

The formulation of the logical structure of the typical, or paradigm, process of evolution, assuming Mendelian heredity, was initially carried out with perhaps greater attention to refuting certain loudly expressed objections to the new outlook than to developing fully its own inherent character. Of the great triumvirate who laid the first foundations for a fully logical, i.e. mathematical, formulation of Mendelian evolution, Haldane and Fisher were English—and England had just seen one of the most ferocious (and silly) of all academic battles, in which the anti-Mendelians, lead by Pearson and Wheldon, had gone down to defeat at the hands of the believers in Mendelism and Discontinuous Variation, whose champion was Bateson[1]. In the heat of the fight there had been some corrosion, on both sides, of trust in Darwinian mechanisms of evolution. Bateson, emphasizing discontinuity not only in heredity but also in phenotypic variation, toyed with ideas of Mutationism allied to those of de Vries, while believers in various types of overriding evolutionary forces, such as orthogenesis, found that the vagueness of the material basis for the hereditary system postulated by Wheldon and Pearson offered scope for their own equally nebulous ideas. Thus the two English progenitors of what came to be called 'Neo-Darwinism' considered that one of their main tasks was to establish, as clearly as possible, that Darwinian natural selection would, after all, 'work' in Mendelian populations. The other main pioneer, the American Sewall Wright, was less affected by these predominantly British squabbles. His mathematical formulation is far less drastically simplified for polemical purposes; but its richer intellectual content calls for subtler and more difficult mathematics, and until recently it has been less influential than those of Haldane and Fisher.

Before attempting to re-formulate the essential logical features of an evolutionary process, it will be as well to remind ourselves of the formulations which were given by the pioneers. The first and simplest of them was Haldane, beginning in a series of papers in the obscurity of the *Proceedings of the Cambridge Philosophical Society* from 1924 onwards. This work was summarized in his book *Causes of Evolution*, 1932. An indication of the atmosphere in which it was written can be found from the quotation he chose as the motto for the Introduction: 'Darwinism is dead.—Any sermon'. On page 20 we find the statement, surprisingly apologetic to modern eyes: 'But I propose to anticipate my future argument to the extent of stating my belief that, in spite of the above criticisms, which are all perfectly valid, natural selection is an important cause of evolution.' The argument, when it comes, is related to systems of which the paradigm case is described as follows (pp. 180, 181): 'In a random-mating group a population composed of the three genotypes in the ratio $u^2AA:2uAa:1aa$ is stable in the absence of selection, and any group whatever reaches this stable equilibrium after a single generation of random mating . . . Now after selection the population $u_n^2AA:2u_nAa:1aa$ is reduced to $u_n^2AA:2u_nAa:(1-k)aa$.'

Haldane developed this paradigm mainly by studying the rates of change of gene frequency, for genes of various kinds, in populations with different types of mating system.

The most drastic simplifications involved in this paradigm are:

1. The system essentially implies an equilibrium, in which the frequency of a gene selected against is reduced to zero, or to the level at which it is maintained by recurrent mutation; but the paradigm assumes that the initial conditions are not at equilibrium. Nothing is explicitly stated about why this should be so; it might be, for instance, because a change in environmental conditions has altered selective values, or because a totally new gene has occurred by mutation.

2. There is no explicit mention of the phenotype, and certainly no hint that phenotypes can be affected by environments as well as by genotypes.

3. There is no mention of the fact that the effect of a given gene is influenced by the rest of the genotype. In some of the later develop-

ments Haldane does discuss specific interactive effects between two or more genes, but he leaves on one side the pervasive effects of 'the genotypic milieu', to use the terminology of that time.

Fisher's paradigm avoids the last of these implications, but otherwise differs more in mathematical technique than in logical structure. Instead of measuring selection coefficients by linear coefficients, such as Haldane's k, he uses 'Malthusian parameters' expressed in exponential terms, which state the numbers of offspring produced by organisms of the genotype in question. In his paradigm, the difference in the Malthusian parameters for two alleles enters an expression which also involves the frequencies of the alleles and a parameter which expresses the phenotypic difference produced by altering one allele to the other in the actual population, taking into account its breeding structure (e.g. amount of inbreeding or assortive mating) and all the other genes present. He develops his paradigm initially into 'The Fundamental Theorem of Natural Selection', in the form: '*The rate of increase in fitness of any organism at any time is equal to its genetic variance in fitness at that time.*'

This is a statement which has proved extremely difficult to interpret. In the first place, it is clear that the word 'organism' must be shorthand for 'population of organisms', but though this makes it easy to attach a meaning to 'genetic variance in fitness', it does nothing to elucidate what may be meant by an 'increase in fitness of an organism (= population)'. It is usually held to imply, if not to be synonymous with, an increase in the numbers of the population; and since genetic variance is essentially a positive quality, this would lead to the conclusion that all animal populations must always increase in numbers, which they do not. Some way has to be found to get around this difficulty. We will return to it later in connection with Maynard Smith's remarks. Here I would only point out that Fisher's paradigm still involves the first two simplifications characteristic of Haldane's.

It is not so easy to disentangle and describe any particular situation as the paradigm adopted by Sewall Wright. He was not concerned to demonstrate the point on which Haldane felt he had to make a bold assertion: 'that natural selection is an important cause of evolution'. He took this for granted, and was more interested in the circumstances in which the operations of natural selection are mitigated or even overcome by other factors. However, his basic picture of an evolving population has a further element of inclusiveness and flexibility over and above that introduced by Fisher. Wright deals in the selective values of whole genotypes, considered as combinations of alleles at large numbers of loci. These values are envisaged in terms of a hyper-surface in a space in which fitness provides one dimension, while the others express the vast number of possible gene combinations. In his earlier papers at least, Wright was mainly concerned with an initial situation whose non-equilibrium character he carefully defines. He conceives of the fitness hyper-surface as comparable to a rough piece of country, with many hills and valleys; and he sets out to consider the mechanisms by which a population, which for some contingent reasons finds itself at the top of one hill, may travel across a valley and thus reach the top of some other, possibly higher, hill in the neighbourhood. Much of his work is therefore concerned with what might be called quantization processes in evolution; and the mathematical tools he uses deal largely with changes in the frequency distribution of gene frequencies in populations, particularly with processes which lead to certain alleles becoming 'fixed' at frequencies of 0 or 100 per cent.

Wright's treatment began by avoiding simplification 1, i.e. failure to specify an initial state of disequilibrium, but only by invoking a rather special case, one in which a population has, by chance, got into a metastable position. His formulation can without too great difficulty be modified to deal with other types of initial non-equilibrium conditions, such as heterogeneity of the environment in space or time, but it remains true that these are not explicitly incorporated into any general paradigm. Further, he makes little more open reference to phenotypes than do Haldane or Fisher, and he does not incorporate into his scheme any suggestion that the phenotypes on which selection acts are affected by environments. He therefore employs simplification 2 just as the others do.

Now, on the face of it, the two great problems of the Theory of Evolution—once we have granted that natural selection is an effective agent—would seem to involve just those points omitted by simplifications 1 and 2. One problem is adaptation, and the focus of the long-continuing debate about Lamarckism is precisely that organisms so often exhibit adaptations which *look as though* they were responses to the environment, but which turn out not to be so in any direct way. Any paradigm which omits the effects of environments in altering phenotypes would seem to make it difficult, if not impossible, to deal with this (leaving it to 'random mutation' is not dealing with it). It was only by taking this factor into account that a solution could be found, in the form of genetic assimilation. Again, the second main problem is that of speciation. Here again everything leads us to the conclusion that diversity of the environment in space and/or time is of the essence, and that a paradigm which implies that the fitness of a genotype is single-valued is likely to prove inadequate.

Maynard Smith's defence of neo-Darwinism. The comments made above do not in any way imply that we should abandon neo-Darwinism (hereafter contracted to neo-D, which I shall also treat as an adjective); they only suggest that some of the simplifications on which the mathematical theory has been based have outlived their usefulness and should be revised. The nature of the problem involved may be better appreciated if one looks at the article by Maynard Smith [2], in which he tries to demonstrate that neo-D is not a mere tautology by enquiring how one might refute it. I shall argue that none of his suggested 'refutations' would really require us to abandon the theory. (As a side-issue, I should like to remark, as I have done earlier [*Ethical Animal*, p. 151], that I have never been convinced by Popper's argument that, while hypotheses cannot be proved, they can be disproved; in practice they can always be suitably amended to deal with the objections raised. Popper encourages a fashionable current of thought in the philosophy of science, which states that the thing to do with a hypothesis is to try to refute it. This is the treatment Maynard applies here, and in my opinion the result suggests that it is not a very useful line of approach—searching for an improved paradigm, as suggested for instance by Kuhn, may prove more rewarding.)

Let us consider, in turn, the various possible 'refutations' which Maynard Smith describes.

The first, which he dismisses undiscussed, would arise if 'we can show that organisms do not multiply', and multiplication he has defined as 'increasing in numbers in at least some environment'. Now it is perhaps a quibble, but it might be pointed out that a once-numerous species which had lost the power of increasing in numbers in any environment might still undergo neo-D evolution for some period while its numbers were declining; it is not unlikely that the Sequoias, for instance, are in this situation. But in general I agree that no serious line of attack is likely to emerge in this connection. I also agree that it would be fatal to neo-D if we could show that organisms do not vary; but it would be fatal to the idea of evolution in general, not only to the neo-D version of it.

Maynard Smith then speaks of 'the assumptions about heredity and the origin of variation', and he has stated that a refutation would occur if 'it could be shown that the assumptions made by neo-D are not in fact true of all organisms'. Now, of course, many different sorts of non-Mendelian heredity have been demonstrated in a variety of organisms—episomes in bacteria; chloroplastal, mitochondrial or more general types of non-chromosomal genes, such as Sager's; organelles in the cortex of Ciliates; and so on. Some types of bacterial transformation, or episomal heredity, could even be interpreted as examples of Lamarckian phenomena. And there are certain rather weak, 'inertial' effects of the kinds he mentions; for instance the occurrence of a duplication of a locus makes possible the evolution of a protein dimer with two related polypeptide chains. If these were really refutations of neo-D it would have been refuted already. But in fact, of course, they are regarded as mere details and special cases. The main body of the theory is not noticeably weakened, though it has to give up the claim—which its more enthusiastic protagonists sometimes announce—to be the sole and sufficient explanation of all evolutionary phenomena.

All these attacks on neo-D are, in fact, attacks simply on Mendelism, not on anything which neo-D has added to Mendelism. To refute neo-D in this manner requires no more, and no less, than a refutation of Mendelism.

Maynard Smith then turns to some suggested refutations based on rates of change, but points out, justly in my view, that we cannot make any quantitative predictions in this field, and that therefore no refutations are possible.

Finally he turns to 'examining the end-products—the existing organisms'. He tries to invent animals whose organs exhibit an order which is clear enough to be undeniable but which it is implausible to attribute to any form of adaptation for reproductive efficiency. I think he fails to realize that he has come into a region where there are also epigenetic rules and types of organization to be considered. For instance, his first example is: 'If someone discovers a deep-sea fish with varying numbers of luminous dots on its tail, the number at any time having the property of being always a prime number, I should regard this as rather strong evidence against neo-D.' I should not draw such a conclusion so quickly. Which prime numbers? If they were 1, 3, 5, 7, 11, 13, 17, 19, I should suppose that the spot-producing mechanisms worked with some threshold-type action, so that at a low level it could produce 1 spot, and at higher levels added more spots two at a time until it got to 7, then the next effective jumps gave an extra 4. We might explain 13 by saying that going from 7 through 9 to 11 put us well above the relevant threshold, and therefore the next jump goes back to being only a 2-jump; and after that, of course, we would get back to a 4-jump and reach 17 (figure 1). Then we would deal with 19 as we did with 13, and 23 as with 17. Then we'd go up another notch to a 6-jump, and get 29. After that I'd be willing to pass it back to Maynard Smith and remind him of his article [3] about epigenetic mechanisms for counting large numbers.

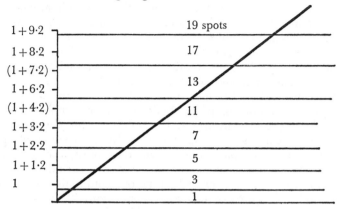

Figure 1. Hypothesis for explaining prime numbers of spots. One spot appears in any case. Further spots are added, two at a time, according to the concentration of a substance produced under the control of the genotype, which varies within the population. Each phenotype has a certain range of stability with regard to this substance, these ranges (shown on the graph) gradually increasing at higher concentrations of the substance. The adjustment of these stability-ranges to get the first eight prime numbers is not wildly implausible.

He goes on: 'And if the dots took up in turn the exact configuration of the various heavenly constellations, I should regard it as an adequate disproof.' If we are to take 'exact' quite literally, this might carry some weight; but if we allowed some latitude in the configurations, the observation, if made, would in my opinion refute not neo-D, but some as yet unformulated theory of epigenetic mechanisms. All neo-D can say about the configuration of spots is that they are useful; there is no *a priori* way of telling whether the fact that they look like something else is relevant or not; if they look like something that seems unlikely to be relevant, the first hypothesis is that the epigenetic system imposes limitations which make it impossible to obtain the useful effects without this associated surprising side-effect.

I should like to give two concrete examples in this connection.

There is a family of Lepidoptera known popularly as 'Lantern Bugs'. Many of the species have large prolongations on the front of the head. In the species *Laternaria lucifera* this prolongation bears a number of spots and patches which give it an extraordinary resemblance to a crocodile's head. Now the prolongation is only a few millimetres long; it is difficult to believe that any animal which has been predated by crocodiles sufficiently to have evolved avoiding mechanisms will ever mistake the moth's head for a real reptile. What adaptive value can have led to its evolution? Is not this almost as odd as Maynard Smith's 'absurd example' of a fish with spots on its tail arranged like the stars of the Great Bear? I cannot explain it, but I do not propose to abandon neo-D on that account.

My other example is, please believe me, not wholly frivolous. When those neo-D stalwarts J.B.S.Haldane and Julian Huxley were young men, they co-operated in writing an extremely good elementary book on *Animal Biology*. For the frontispiece they went, like Maynard Smith, to the fauna of the deep sea. And they picked out illustrations of two species which no one could fail to see as very recognizable caricatures of the two authors (figure 2). Now Haldane and Huxley could certainly be formidable adversaries to run across unexpectedly in the darkness of some deep argument; but no, that cannot have been the neo-D reason for the evolution of these fish. The resemblances are as unexplained as would be the constellation-spots. One can say little more than that, if enough species go on evolving long enough, a lot of funny things are likely to turn up.

The point that is non-frivolous about this example is that the deep-sea fauna is one of those which presents in dramatic form one of the great problems of evolution. How did it come about that any animals ever went into such an uninviting and difficult environment? The standard neo-D paradigm, particularly in the Fisher-Haldane version, has altogether too little to say about the colonization of new habitats, especially of habitats which are poorer than the original home of the species, as the abysses are poorer than other parts of the sea.

Towards a post-neo-Darwinian paradigm. As a matter of fact, Maynard Smith actually gave away the whole story about 'refuting' neo-D before he started to show how he thought this might be done. On page 83 he admits that the phrase 'the survival of the fittest' *is* a tautology (and therefore irrefutable), if the word fitness is used in its neo-D sense. He claims that it should be used, in that phrase, in some other sense: he suggests 'adaptive complexity', but does not discuss this much further, except for the statement that complexity

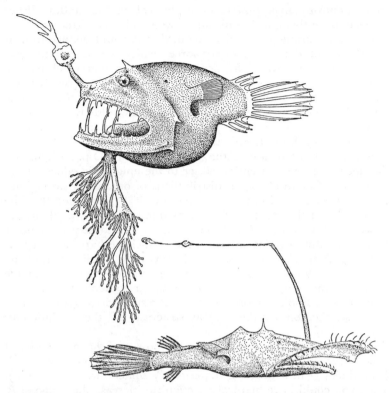

Figure 2. Two deep-sea angler-fishes, to show their extraordinary structure and their adaptations to their mode of life.

cannot be precisely defined. One way of describing the aim of a search for a new paradigm would be to say that it is an attempt to define more fully what this concept of 'fitness' should be. It is, after all, very generally accepted that fitness is a difficult and obscure concept. The simple Haldane-Fisher fitness has to be modified in situations which are technically more complicated, e.g. with non-random mating, or selection intensities dependent on gene frequencies[4]. More radical modifications have been made by people studying heterogeneous environments or more complex genetic systems[5].

A few authors have tried to formulate a concept of a general parameter which will always change in one direction during evolution, as entropy always increases in physical systems, and as Fisher seems to have thought that his fitness would always increase.

If such a parameter could be defined, one could deduce from it the nature of the 'evolutionary force' which keeps evolutionary processes on the move in the face of the many factors which tend to bring it to a halt at some position of equilibrium (see below). Among possible parameters, MacArthur[6] has suggested that one measuring efficiency of using limited resources would always increase in evolution; Slobotkin[7] proposes that homeostasis is always increased. But both these authors, in my opinion, began their arguments from too close to the conventional neo-D paradigm, and did not pay enough attention either to the epigenetic effects of the environment or to changing distributions of organisms throughout a heterogeneous environment. I have myself[8] suggested that the parameter which is continually increased in evolution would have to express ability of members of the system to find some way or other of keeping alive and leaving offspring. This is perhaps not very far from MacArthur's suggestion, but involves the possibility that the organisms may not simply become more efficient at using available resources, but may begin exploiting new resources. The way to give more precision or more penetration to such ideas is, I suggest, a new investigation of the logical structure of the evolutionary process.

The post-neo-Darwinian paradigm. 1. Suppose you have a material structure P with a characteristic Q such that the presence of P with Q produces Q in a range of materials P_i under circumstances E_j. Then you could have 'natural selection' to increase the range of P_i and E_j. (For instance, a crystal dislocation Q in material P, where Q would be replicated in a variety of materials P_i in several environments E_j.) But the whole system would not qualify as 'life' because, in my original phrase, it is not interesting enough: in rather more objective language, because this set of postulates makes no provision that Q has an effect on E. But Q can only affect E if it is not merely a memory-store to be replicated, but is also an operator. To say that Q becomes an operator is the same as saying that Q becomes a phenotype. This might, logically, involve no, or very little, or very much, translation of the memory-store Q into the effective operator phenotype Q^*. But in practice—and perhaps because of a profound law of action-reaction—it is difficult (impossible?) to find a Q which is stable enough to be an efficient store and at the same time reactive enough to be an effective operator. Thus a considerable translation from Q to Q^* is characteristic, and may be necessary, for all systems which can be accorded the name of 'living'.

2. Having now got a system in which natural selection will evolve something effective (or interesting) enough to be called 'life', we need to specify conditions under which evolution will continue. If the range of conditions E_j is single-valued, evolution will produce a single optimum phenotype Q^* (probably, but not necessarily, depending on a single optimum genotype Q) and then stop. Similarly, if E_j has only a finite number of values, evolution will eventually reach a stable end-state, with a number of different Q^*s, each optimum for one value of E_j. The paradigm situation therefore demands two further conditions: (a) that E_j is an infinite-numbered set; and (b) that there are sufficient Qs to provide Q^*s suitable for an infinite sub-set of E_js.

The infinity of E_js is ensured by the fact that Q^*s are components of E_js. Thus any evolution of a particular Q^* into Q'^* automatically changes a number of E_js in which this Q^* is a component. In more biological language, the environment which exerts selection on one organism is influenced by the presence of other organisms; and as the other organisms change in evolution, so the environment of the first organism is altered, and it must evolve too. In a still cruder formulation, we can say that the evolutionary appearance of a new species automatically creates potential new environmental niches ready for exploitation by some further new form. ('Big fleas have little fleas, upon their back to bite 'em, and little fleas have lesser fleas, and so *ad infinitum*.') [9]

The second requirement, that the available genotypes must be capable of producing phenotypes which can exploit the new environments, requires some special provision of a means of creating new genetic variation. It might at first sight seem simplest to ensure this by a mechanism in which the new environment would itself stimulate the production of new appropriate hereditary variation. This is the Lamarckian hypothesis. It is indeed a fact that new environments can often produce new appropriate phenotypes (E'_j produces Q'^*). But except in very special circumstances E'_j does *not* produce Q'. Instead, what we normally encounter are systems in which Q is continually giving rise to a range of altered forms (Q'_1, Q'_2 ...) by a process which is 'random' in the sense that it is unrelated to any selectively-effective E_j (though it must certainly be controlled by rules of occurrence not related to some E_j).

The relative importance of these two 'evolutionary forces'—new environments, new genes—needs some consideration. The classical neo-D position is that all gene mutations occur with a specifiable frequency, so that there is no question of genuinely novel, first-time-ever mutated genes: and that mutation frequencies are so low (10^{-5} or less) that they can almost never overcome the selection pressures. Such considerations give rise to a dogma of the ineffectiveness of mutation as an evolutionary force; it merely provides some raw material on which other more effective forces might act. But this dogma grew up in the pre-molecular period of genetics, when one was thinking of, e.g. the mutation rate of the plus allele of the 'white eye locus' in *Drosophila* to an allele which gave a white eye phenotype. We have to rethink the situation now that we realize that the protein produced by the w locus (or group of proteins, from a group of loci [10] — but that is an irrelevant complication from the present point of view) can be reduced to complete ineffectiveness by changes in any one of a large number of different peptides. If one characterizes mutations, not by such crude criteria as 'producing white-eyed rather than red-eyed phenotypes', but in molecular terms as substituting amino-acid P for Q at position W in a polypeptide, then the rate of particular mutational steps is reduced, probably by several orders of magnitude. This means that they are even less able to overcome the effects of selection by sheer frequency of occurrence; but it also raises the possibility that some mutation rates may be so low that in comparison with the time-scale of the relevant evolutionary events, it may not be justifiable to consider them in terms of continuous rates at all. We may have to bring back into our thinking the possibility of radically new, never-seen-before, mutations — considerations which were in the forefront for the earliest Mendelian evolutionists, such as Bateson and de Vries in the first decade or so of this century.

   Before one can evaluate the relative importance in evolution of (i) changes in environment, as against (ii) mutations-as-alterations-of-molecules or mutations-as-determinants-of-epigenetically-complex-phenotypes, one needs to realize the range of spectrum of levels of organization in which evolution may occur. At one end of the spectrum is:

(a) the purely macromolecular. Is it (selection-wise) a 'good thing' to substitute valine for glutamic acid at position 6 from one end of the chain of haemoglobin (producing the sickle-cell character)? In higher organisms there probably is a gene for doing this floating around in the population anyway — or you could find some other way of dealing with the environmental selective pressure. The question is more important in relation to the very earliest phases of evolution, when the living systems comprised little more than a bare minimum of molecules which could just keep going. How did one improve the very first, barely effective, DNA polymerases, oxidative enzymes, or chlorophylls? The course followed by evolution may, at such stages, really have been influenced by the nature of the polypeptide substitutions that mutation threw up.

(b) From this, there is a complete continuous spectrum of increasing levels of complexity—through the 'hypermolecular', which is what classical neo-D concerns itself with—to the 'ecosystem' evolution involved in such a question as: how did the London sparrow cope with the success of the petrol engine in driving off the streets all those horses, with their offerings of dung full of delicious seeds? How did the rabbit of the gentle English fields evolve a method of flourishing in the harsh Australian outback, with not a single species of its normal food plants, or the red deer of the Scottish heather-covered hills find its evolutionary way into a set-up in which, ecologically associated with the Australian wombat, it dominates much of the wilder country of New Zealand and, incidentally, grows to about twice its normal weight in its home country?[11]. It is extremely unlikely that any of these evolutionary episodes had to wait for any new polypeptide sequences. All the Qs necessary to produce the appropriate Q^*s were, one imagines, already present in the population, and the effective 'evolutionary force' was the occurrence of the new environments (E_js).

The relative importance of the two 'evolutionary forces' arising from new random mutations and new environments must therefore change as we pass from the macromolecular end of the evolutionary spectrum through the hypermolecular to the ecosystem end, but wherever we are within this range, each of these 'forces' has *some* importance.

It is important to emphasize that the new genetic variation must not only be novel, but must include variations which make possible the exploitation of environments which the population previously did not utilize. In Longuet-Higgins' terminology, it is not sufficient to produce new mutations which merely insert new parameters into existing programmes; they must actually be able to rewrite the programme. Essentially the same point was made by Pattee. In a letter (12 February 1968) discussing tactic copolymer growth as an elementary genetic system (see also [12]) he writes: 'The fundamental reason that evolution is limited in such single tactic co-polymers is that the rule for monomer addition or the code is intimately dependent on some fixed length of the description. If evolution is to lead to unlimited complexity, then the code must accept a description of indefinite length and the description itself must be able to grow indefinitely without changing the code.'

The predominance of 'random mutation' over Lamarckian mechanisms in tellurian ('this-earthly') biological systems does not arise because the latter could not, in theory, support evolutionary processes. It is presumably related to the problem described at the end of paragraph 1, how to combine a store which is unreactive enough to be reliable, with something which interacts with the environment sufficiently actively to be 'interesting'. To ensure reliability the store must be rather unreactive; the solution which has been generally adopted by tellurian systems is to provide the necessary variability by a random process not depending on reaction with the environment rather by endowing the storage material with some residual reactivity [13].

3. We need also to provide that the variant Qs actually come in contact with the variety of new E's. This may be brought about simply by arranging for very wide but quite indiscriminate geographical dispersion of newly fertilized zygotes; and this is the mechanism on which the whole plant kindgom relies. The world of non-sessile, motile animals has a more economical method of achieving this result by employing behavioural mechanisms, which lead animals either to explore situations more or less at random (which is little better than plants can do) or, more typically, to choose those environments in which they can most easily earn their daily bread and escape their enemies. This is an elaboration, of a rather Lamarckian character in that it involves a reaction to the immediate circumstances, which is over and above the absolute necessities of the evolutionary paradigm; but one which is very important in animal evolution.

4. In order to accommodate the numbers of altered or mutated Qs required to evolve into the new environments E'_j, the paradigm situation must be one describable only in terms of populations, not of individuals.

5. The new environments E'_j are, as we have seen, essentially complex entities, being functions of a number of variables, among which are, for instance, a number of phenotypes Q^*s of various species, as well as various physical quantities, etc. The evolution of an optimum phenotype for a particular new environment, $Q^*_{E'_j}$ will therefore in many cases involve combination of a number of different variant genotypes, Q'_p, Q'_p, Q'_r, . . ., etc. It is therefore not surprising to find that most evolving systems have developed some method of encouraging the production of appropriate combinations of Qs. Provision for this is not an absolutely necessary

constituent of the evolutionary paradigm, since some organisms get away with nothing more radical than very large numbers of short life-cycles, though there are few, even of the viruses and bacteria, who cannot do anything better than this. Nearly all organisms have evolved mechanisms which facilitate recombination of variant Qs, usually by some sexual or parasexual process.

6. In so far as the necessary heterogeneity of the environment comes into existence as a consequence of interactions between existing organisms (i.e. the E'_js are functions of $Q's$), two consequences will follow: (a) There will be a continual increase in the number of species which are optimum in some or other of the ever-increasing number of E'_j. Thus the diversity of the organic world will continually expand; (b) Since a new, attainable E_j may be a function of an indefinite number of existing phenotypes, there will be a tendency for the evolution of ever more complex phenotypes, capable of operating optimally in environments of increasing complexity. When there are both water-snails and water-visiting mammals, evolution can produce parasites with life-cycles involving interactions with both hosts; when there are night-flying insects, then, but not before, evolution can produce larger night-flying predators, such as bats, which need a sonar system to prevent them running into obstacles or to locate their prey.

7. The necessary heterogeneity of environment has another consequence, which cannot be omitted from the paradigm situation. Selection operates on phenotypes, and phenotypes are affected by environments as well as by genotypes. Further, since there is a necessity for mobility (whether passive or active) to ensure that new environments are explored, it will not in general be true that the environmental influences which contribute to the formation of the phenotype are identical with those which exert the most important selection on it. (Consider, for instance, an annual plant; the phenotype of a plant and its seeds will be influenced by the weather, etc., in the year n, while the main selection may be exerted by the weather in year $n+1$, when the seeds germinate.) Thus the paradigm situation must incorporate two theoretically separable effects of any environment E_j: on the development of a Q into a $Q^*_{E'_j}$, and on the selective values of the variety of Q^*_x which arrive within or pass through it.

The complete paradigm must therefore include the following items: a genetic system whose items (Qs) are not mere information, but are algorithms or programmes which produce phenotypes (Q^*s); there must be a mechanism for producing an indefinite variety of new $Q'*$s, some of which must act in a radical way which can be described as 'rewriting the programme'; there must also be an indefinite number of environments, and this is assured by the fact that the evolving phenotypes are components of environments for their own or other species. Further, some at least of the species in the evolving biosystem must have means of dispersal, passive or active, which will bring them into contact with the new environments (under these circumstances, other species may have the new environments brought to them). These environments will not only exert selective pressure on the phenotypes, but will also act as items in programmes, modifying the epigenetic processes with which the Qs become worked out into $Q*$s.

From the standpoint of this paradigm, how should we envisage the classical evolutionary concepts of adaptation and fitness? I confess I have not sufficient mastery of the sophisticated mathematics which would be necessary for a rigorous exposition. Instead of attempting that, I will suggest, for adaptation, an allegorical illustration which I originally applied to man in his world (motto to chap. 4 of *Behind Appearances*), but which I think applies almost equally well to any organism: *Man in the world is like a caterpillar weaving its cocoon. The cocoon is made of threads extruded by the caterpillar itself, and is woven to a shape in which the caterpillar fits comfortably. But it also has to be fitted to the thorny twigs—the external world—which supports it. A puppy going to sleep on a stony beach—a 'joggle-fit', the puppy wriggles some stones out of the way, and curves himself in between those too heavy to shift—that is the operational method of science (and of the evolution of biological systems).*

As regards fitness, we have to define that concept in terms which allow for the existence of heterogeneous and evolving environments, and of organisms capable of active or passive dispersal through a range of environments which are acting both as agents of selection and as subsidiary programmes affecting the development of the phenotypes which are selected. The fitness even of a single phenotype cannot, therefore, be represented by a single-valued coefficient, but only by a matrix, or a continuous distribution of values, which specifies also the variety of environments, in which selection may occur. If we wish to attach 'fitness' to a single, multigenic *genotype*, we should have to increase the dimensionality of the matrix so that it could take account also of the epigenetic-programming aspects of the environments; and if we wished to emulate the classical neo-Darwinists and speak of the fitness of single genes, we should have to increase the matrix again to incorporate all the various genetic combinations in which it might occur in the population in question (it seems highly dubious to me that any process of averaging over all these combinations, as advocated, e.g. by Fisher, has any biological validity).

In contrast to the difficulties of conceiving of 'the fitness of a gene' from this point of view, the concept of the fitness of a population comes close to giving one an opportunity to use the adjective 'perspicuous' in the precise sense given to it by that pedant Fowler (*Modern English Usage*)—'means (the being) easy to get a clear idea of'. The fitness of a population is the degree to which its gene pool gives it the ability to find some way or other of leaving offspring in the temporarily and spatially heterogeneous range of environments which its dispersion mechanisms offer to it.

But, even if not too difficult to formulate, this concept has a joker in it—which is what makes it interesting and challenging. How to know what the 'temporarily heterogeneous range of environments' may bring? A new Ice Age, a new virus, a new predator? The gene pool can, of course, preserve for some time genes which turned out to be useful in relation to critical situations in the more or less recent past: Lewontin [14] has discussed these possibilities in a very stimulating way, and shown that, unless the environment behaves itself—is not capricious—they are rather limited. From a more general point of view, one can envisage a number of possible strategies. As (a) to canalize your development, build up well-buffered chreods, and insist on developing into some good all-purpose almost invariant form in spite of whatever environmental effects get thrown into the epigenetic programmes—as mice concede little more than a fractional elongation of the tail to the difference between growing up in a hothouse rather than a cold-storage depot [15]; or (b) acquiring a gene pool which allows an extreme flexibility in the end-results of development—good examples are some small crustacea, such as *Daphnia* and *Artemia*, in which the minute physico-chemical variations between every pond of water are reflected in the shape of the adults, or plants such as the 'water arrowroot' *Sagittaria sagittifolia*, in which the leaves of one and the same plant have radically different forms when they are growing wholly under water, on the surface, or in the air. To make a success of this gambit, it is of course, necessary to ensure that the plasticity of epigenesis is in general, or at least in really crucial situations, such that the environmentally-modified forms are selectively useful in the environments that produced them. It is no use allowing your muscular development to be influenced by the environmental demands on the use of muscles if the dependence takes the form that using muscles causes them to be consumed and to wither away; (c) another ploy is to develop some general defence mechanisms which do not have to know in advance what they will have to defend against. 'Random' mutation provides this facility to some extent, but only over periods of many generations of selection. If the population can afford to wait that long, as bacteria can, to add a general mechanism for rapidly spreading a good defence, once it has been acquired, by such systems as Infective Resistance Factors, is obviously a useful second stage. Perhaps the best example of a generalized defence is the vertebrate antibody-production system, which seems able, even within one lifetime, to protect against an enormous range of invading foreign substances.

The systematic exploration of the evolutionary strategies in facing an unknown, but usually not wholly unforecastable, future would take us into a realm of thought which is most challenging and very characteristic of the basic problems of biology. The main issue in evolution is how populations deal with unknown futures; is this problem so different from that described by Gregory, when he says that 'perception involves the continual solution of a series of puzzles'? In epigenesis we find systems which will develop into perfectly good lenses or livers, even when there is something non-standard in the conditions which normally guide the cell's synthetic machinery into those paths. In all these cases we are forced to consider the nature of mechanisms which can operate effectively on the basis of inadequate information. This seems to be one of the central general problems of Theoretical Biology. Life might be defined as the art of getting away with it; and Theoretical Biology as the attempt painstakingly to explicate just how it is done.

SOME COMMENTS ON WADDINGTON'S PARADIGM
BY J.MAYNARD SMITH

1. *Neo-Darwinism and Mendelism.* Wad argues that 'to refute neo-D requires no more, and no less, than a refutation of Mendelism'. This of course depends on how you define neo-D. In my 'defence' of neo-D I made Weismannism rather than Mendelism the central assumption. Otherwise the refutation of neo-D is trivial: bacteria do not Mendelise but they do evolve. But the Weismannist assumption (roughly, if the phenotype of an individual is altered by an altered environment, this will not cause that individual to produce offspring with the new phenotype) is not disproved by the types of non-Mendelian heredity mentioned by Wad.

Nevertheless, I agree with the main point Wad is making—the most direct way (but not the only way—see below) of refuting neo-D is to show that its genetic assumptions are wrong.

2. *Refutation by the end-products of evolution.* I argued that complex structures which did not contribute to the survival of their possessors would refute neo-D. Wad argues that since development is based on algorithms, it can lead to inexplicably complicated (= funny) results. I think that I am right, but I agree that no single example could be decisive. My point is that when biologists are confronted by a structure, they analyse its functions in terms of its contribution to survival, and this method of analysis usually works. If it didn't— i.e. if it often turned out that an organ when studied in detail could not be interpreted as contributing to survival—then biologists would have abandoned Darwinism long ago. But it is of course true that there are at any one time plenty of structures, usually ones which have been little studied, which cannot be interpreted adaptively.

3. '*The survival of the fittest*' *and Waddington's paradigm.* It is perhaps a pity that this phrase was ever introduced into biology, because it is a standing invitation to philosophers to argue that Darwinism is a tautology. There seems to be three ways of treating the phrase:

(a) Assume 'fittest' means 'most likely to survive', and you have a boring tautology.

(b) Replace 'fittest' by some more sophisticated definition of survival capacity, and you may have an interesting tautology. Thus a conclusion may follow necessarily from certain assumptions, but still be interesting, e.g. the conclusion that planetary orbits are elliptical, given Newton's laws of motion and gravitation. This is particularly likely to be the case if the conclusions are more easily tested than the assumptions, as is the case in the Newtonian example.

Thus one might try to deduce from the laws of heredity some property which will be maximized. This is what Fisher's 'fundamental theorem' does, and what Wad does in his article. So far the approach seems to me to have been unfruitful, mainly because the assumptions (i.e. of genetics) are easier to test than the conclusions (i.e. the course of evolution).

If Fisher's 'fundamental theorem' is interpreted to mean that evolution must lead to an increase in the rate of growth in numbers (i.e. 'fitness') of a population, then the theorem is simply false. If, alternatively, one interprets fitness as a mathematical function of the frequencies of genotypes in a population and of their *relative* probabilities of survival, then the theorem is, with certain qualifications, true, but, as Wad points out, difficult to apply.

The snag with Wad's less mathematical attempt to find something which increases in evolution is that it leads to a false conclusion. Thus he concludes 'the fitness of a population is the degree to which its gene pool gives it the ability to find some way or other of leaving offspring in the temporarily or spatially heterogeneous range of environments which its dispersion mechanisms offer to it'. Now fitness in this sense is not necessarily maximized. For example, plant species commonly and animal species occasionally lose the capacity for sexual reproduction. Such a change usually leads to extinction. It is a lowering of fitness in Wad's sense, yet I see no reason to doubt that the change occurs by natural selection.

(c) The third approach, which I adopted in my 'defence', is to reformulate the phrase 'the survival of the fittest' in a non-tautological way, by taking fitness to refer not to some function, sophisticated or otherwise, of survival capacity, but to the properties of 'adaptive complexity' or 'harmoniousness' or what have you—i.e. to those properties of living organisms, and sometimes of their artifacts, which distinguish them from inanimate matter, and which call for an explanation.

I do not think approaches (b) and (c) are mutually exclusive (although obviously confusion arises if 'fitness' is used in two senses). The difference between Wad and myself lies not so much in our views about the mechanism of evolution, which are rather similar, but in what we were trying to do. In my article on 'the status of neo-Darwinism' I was trying to defend the present orthodoxy from criticisms of a philosophical and fundamentally Lamarckist type. Wad, perhaps rightly, regards this argument as no longer very interesting, or as something for molecular biologists to worry about, and has therefore been trying to say something new about evolution.

Reply by C. H. Waddington

To my way of thinking, John's comments introduce some confusions into this discussion, which it may be well to try to clear up. They are largely terminological, and connected with the fact that when I offer a criticism of neo-D, John tends to reply with an impassioned defence of Darwinism or Weismannism or some other well-accepted historical precursor of the views I was discussing. Let us first, then, agree on what we mean, at least roughly, by the doctrines attached to these various names.

By *Darwinism*, I mean the theory that organisms come into existence by a process which involves material heredity from their progenitors under the control of natural selection. This is certainly *not* a tautologous statement, since there is an alternative to it— which was in fact generally accepted before Darwin, namely that organisms are brought into being by 'Special Creation', or something of the kind. What John refers to as 'refutation by the end-products of evolution' would be refutation of Darwinism itself, not merely of neo-D.

By *Weismannism*, I mean the same as John does, according to the statement in his first paragraph. This again certainly is *not* tautologous: it is a necessary but not a sufficient condition that evolution should be of the neo-D type. I am not attacking it. Perhaps John is right in thinking that it still needs defending in any company in which philosophers are present, but I don't think we need waste much time on rehashing the old arguments in this meeting.

By *Mendelism* I mean the theory that heredity is transmitted in the form of discrete factors which can segregate and recombine. I think it is confusing to introduce the verb 'to Mendelise', a piece of lab jargon dating from the days of Bateson and Punnett, when it had the meaning 'to exhibit the phenomenon of segregation into classes with one or other of the classical "Mendelian ratios", i.e. 3 : 1, 9 : 3 : 3 : 1, etc.' Many types of organisms (e.g. polyploids) do not Mendelise in that sense, but no one would deny that they exhibit Mendelian heredity. I think the same is true even of bacteria, and do not accept John's contention that their behaviour refutes Mendelism.

By *neo-D* I mean the view that Weismann's doctrine—that there is no influence of the phenotype on the genotype—can be transferred from the individual level to the population level, and that an adequate theory of evolution can be formulated in which 'fitnesses' are attributed to genotypes. John slides altogether too easily between the Weismannist point that the environment of an individual does not affect the heredity he transmits, and the quite different argument that the environment of a population does not affect what they transmit. I maintain that a population's environment does influence, quantitatively, what they transmit, because natural selection acts on phenotypes which are partially environment-dependent. John also gives away too much in his para. 3(a), where he suggests that if we define fitness as 'most likely to survive' we have only a boring tautology, and that we need to define it in a more sophisticated way to get a tautology as interesting as Newtonian mechanics. I have never wished to deny that the results obtained by classical neo-D were as interesting and valuable in their field as Newton's conclusion that planetary orbits are elliptical—and just as unavailable to commonsense unassisted by algebra. The point I am making is comparable to the criticism which might be offered against Newtonian mechanics—that it deals with point-masses, frictionless surfaces, an unresisting medium. It is good mathematics, but as science it is good in outer space, but poor in the sticky conditions on earth. Now, as soon as we have phenotypes we are in a realm whose correlate in the physical world would be that of friction, turbulence, bodies occupying volumes, and all the other complexities which have to be added to the Newtonian picture before it can be actually used. This is what we now need to do to neo-D.

But we do not, in my opinion, need to give up the basic point that fitness is essentially 'survival capacity' (i.e. capacity to leave offspring). I was surprised to find John, in his para. 3(c) ready to allow this position to be overrun by the enemy. But it is the fundamental strategic strong-point of the whole of Darwinism. Once you concede that the thing that survives, i.e. that contributes most to evolution, is something other than the thing which leaves most offspring, then you might as well go straight back to Special Creation and have done with it. The point is not to compromise on the issue that the only way to contribute to evolution is to leave offspring, but to ask more sophisticated questions about just which organisms *do* leave more offspring.

Finally, I do not follow the last paragraph in John's section 3(b). Of course some plant and animal species sometimes have lost the capacity for certain types of reproductive performance, including sexual reproduction in general; but I see no reason to doubt that this loss is produced by natural selection of those individuals with the greatest capacity to leave offspring in the environments immediately available. Of course, again, selection, which operates on the differences in fitness between contemporary individuals, may push a population into an evolutionary situation in which it is unable to cope with changes in its environment, and so becomes extinct. But to state that a certain property is being maximized within a population does not imply that it is always getting greater. Natural selection will pull up fitness as far as it can, but that may still not be far enough to ensure the survival of the population. The point which John is making here against my concept of 'fitness' is the same as that which he advanced to David Bohm against Wynne-Edwards concept; and I feel justified in defending myself even more strongly than I defended Wynne-Edwards ([2], pp. 90, 95).   [1969]

26

A CATASTROPHE THEORY OF EVOLUTION

In this paper I shall discuss mathematical theories of evolution, but I shall not be able to offer any fully developed treatment of a new mathematical theorem. As is well known to Dr Okan Gurel, who invited me to participate, and to anyone acquainted with my work, I am not a mathematician but a biologist. When confronted with a theoretical paper involving mathematics, it has been my practice to take the mathematical developments for granted; other technically qualified experts will pull the author up short if he has been guilty of technical lapses. What interests me is the type of thinking implied by the manner in which the very first equations are formulated. How adequate is it as an analysis of the observational data that we are trying to understand? I shall argue that although the mathematical theory of evolution is probably the most fully developed application of mathematics to any field of biology, the underlying assumptions of present treatments no longer seem adequately to take into account factors that have come to appear of crucial importance. The time has come for a new shift of 'paradigm', to use the word popularized by Thomas Kuhn.

THE PARADIGMS OF MATHEMATICAL EVOLUTION THEORY

It will be well to begin with a sketch of the presuppositions that have underlain the development of mathematical theories of evolution in recent times.

At the beginning of this century the dominant theory of heredity was phrased in statistical terms. Galton, Pearson, Wheldon, and their followers were concerned with the correlations between measurements made on groups of genetically related individuals in a population. The rediscovery of Mendelism led, after very considerable battles, to the adoption of a new paradigm, centred around the study of the behaviour of discrete hereditary units, or genes, in crosses between pairs of identified individuals. In so far as the early pioneers of Mendelism discussed evolution at all, they tended to attribute evolutionary change to the appearance, by random mutation, of new genes that had not been present earlier.

Evolution, however, is a phenomenon that does not occur in individuals—they die—but in populations. A more thorough treatment of Mendelian evolution necessitated a shift of interest from the individual mating to the process of gene transmission within populations. The first success was the demonstration of the so-called Hardy-

Weinberg Law, that if the frequencies within a population of two different alleles of a locus, A and a, are in the ratio $A/a=u$, then if mating is at random, the frequencies of the diploid genotypes AA, Aa and aa, will within one generation reach the equilibrium proportions $u^2\ AA:2u\ Aa:1\ aa$. In the absence of any differential selection selection between the genotypes, they will remain in those proportions.

Haldane was the first to expand this theory, so that it could deal with evolution. He did so by attaching, to the frequencies of the genotypes, coefficients that indicated the relative number of offspring they contribute to the next generation; that is, the intensity of natural selection operating on them. Thus if selection acts against the double recessive, the proportions in the next generation would be $u^2\ AA:2u\ Aa:(1-k)aa$. The course of evolution could then be followed by equations specifying the change in u as generations pass. Fisher developed a similar theory, using an exponential expression rather than a simple arithmetical one for the selection coefficient. At about the same time Sewall Wright developed a different mathematical way of expressing the genetic relations between successive generations, involving the use of path coefficients; and he, like Haldane and Fisher, discussed evolutionary processes by means of coefficients of fitness attached to the genotypes, and reached expressions for the rate of genetic changes under selection essentially similar to those of Fisher and Haldane.

The most important problem areas that have been approached in recent years by these methods are, perhaps, three in number: the consequences of statistically inadequate sampling of populations; the effects of genetic linkage between factors; and the very important point, which has been emphasized particularly by the experimental observations of Dobzhansky, that natural populations are genetically much more heterogeneous than had been thought, and that evolution should be considered not so much a consequence of changes in frequency of single identifiable genes, but rather as resulting from alterations of the proportions in which many different genes are present in the gene pool of a population. A very high degree of algebraic sophistication has had to be deployed to deal with these problems; indeed, analytical methods have broken down in many contexts, and resort has had to be made to computer simulations.

All these mathematical developments, illuminating as they have been, and carried to degrees of refinement where I am quite unequipped to follow them, have been within a paradigm based on the assignment of selection coefficients, or fitnesses, to genotypes. It is this paradigm itself, rather than any theorem within it, which I, as a developmental biologist concerned with how genotypes become translated into phenotypes, have been calling in question for the last thirty years or so. Selection does not impinge directly on genotypes, but on phenotypes. If a horse is escaping from a tiger by running away, neither the tiger—nor anyone else—is interested in its genotype. The question at issue is: how fast and how far can it run? The answer is, of course, influenced by its genotype, but also influenced by environmental factors that have operated during the horse's development; has it been optimally fed? was it trained, by practice (or even perhaps by a racehorse trainer), to get the best results out of the balance between sprint and endurance? and so on. If the horse escapes, and lives to breed, the contribution it will make to the next generation will, of course, be only its genotype; but it is not only on its genotype that its contribution depends.

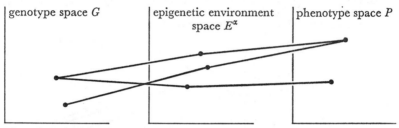

Figure 1. Transition from genotype to phenotype.

The relation between genotype and fitness is far more complex, and I think much more challenging and interesting to the mathematician, than anything contemplated in the Neo-Darwinist paradigm, which merely attaches fitness coefficients—whether arithmetic or exponential—to genotypes. One could begin to discuss it, in simple terms, in some such way as the following. The population of organisms will contain a highly heterogeneous gene pool, each individual genotype being a sample drawn from this. Thus we have to deal with a multidimensional genotype space. The population will inhabit a heterogeneous environment; some horses will grow up on lusher pastures than others, and so on. We therefore have a multidimensional space of 'epigenetic environments'; i.e. environments that affect individuals during their development. The phenotype is the 'product'—I am using this word loosely, I am not sure with what degree of mathematical rigour—of these two (figure 1).

The phenotype is then acted on by a heterogeneous 'selective environment'—heterogeneous because some horses may be lucky enough never to get hunted by a tiger. Note that this multidimensional selective environment space is not the same as the epigenetic environment space; to give a simple example, the epigenetic environment of an annual plant is the weather, etc., this year, which conditions how many seeds it forms; the selective environment, which conditions how many of those seeds actually contribute to the next generation, is the weather, etc., next year. Now, fitness is the 'product' of the phenotype space and the selective-environment space. But it is of much lesser dimensionality. Usually it is taken as one-dimensional, a simple matter of 'how many offspring are contributed to the next generation'. This may be an oversimplification: evolutionary fitness really involves the contribution not just to the next generation, but to an indefinitely prolonged succession of generations, stretching into an unforeseeable future. But let us neglect that complexity for the moment. If we do so, we have a situation as in figure 2.

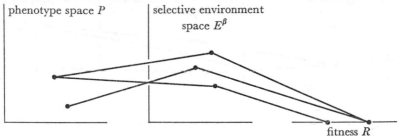

Figure 2. Phenotype and fitness.

It is at this point that the simple biologist, like myself, begins to be acutely conscious of his need for help from his mathematical colleagues in clarifying the logical consequences of such a scheme. As I understand it, the genotype-to-phenotype transition diagrammed is figure 1 could be written as a function $g: E^\alpha \times G \to P$, where E^α, G, and P stand for Epigenetic Environment, Genotype and Phenotype; all multidimensional. Then the phenotype-to-fitness relation, in figure 2, could be written $f: E^\beta \times P \to R$, where E^β, P, and R stand for Selective Environment, Phenotype, and Fitness; with Fitness one-dimensional. One could then put these together into a map

$$E^\alpha \times G \to E^\beta \times P \to R$$
$$\underbrace{\qquad\qquad}_{\phi}\nearrow$$

The first problem I wish to raise is: what are the consequences of the fact that R is of much less dimensionality than G, P, or either of the Es? Does this not imply that there can be considerable variation of phenotype without any change in fitness? That is, there must be points in $E^\beta \times P$ for which $\mathrm{grad}_P f = 0$. This in turn implies that there are points in $E^\alpha \times G$ for which $\mathrm{grad}_\phi = 0$, and, more significantly, there are fold points of G where $\mathrm{grad}_\phi = 0$ but $\mathrm{grad}\, f \neq 0$.

But if so, we find ourselves dealing with a system of controls (operating by selection) in which response is not a continuous function of the controlling variables. The system involves discontinuities—thresholds, in the older terminology of biologists, 'catastrophes' in the language used by mathematicians such as Rene Thom [1] and Christopher Zeeman [2]. To give a simple biological example: if your phenotype is such that you cannot make much of a living as a professional heavyweight boxer, you may—in the multidimensional selective environment man inhabits—get by, to leave just the same number of offspring, by becoming a watch-repairer or a bank clerk or a computer programmer. Your fitness will change only if you cannot find any selective environment in which you can make a go of it.

The usual convention is to say that natural selection tends to maximize fitness. As I shall want to talk about stabilities of fitness, it may be more convenient to reverse the signs, and say that selection tends to minimize the misfit between phenotype and environment.

There is another relation that engenders catastrophes in the evolutionary process. This is in the $g: E^\alpha \times G \to P$ function. It is an empirical fact of observation that phenotypes are remarkably invarient with respect to alterations either of genotype or environment. It is quite difficult to provide theoretical reasons why this should be so, but I do not intend to embark on that problem now. Let us accept that, in actuality, many changes of genotype do not come to phenotypic expression. Names, but not explanations, of the phenomena involved are such expressions as 'dominance and recessitivity of alleles', 'epistasy of gene loci', and the like. Similarly, many even quite drastic environmental influences on a developing individual turn out eventually to bring about no alteration of its phenotype; the embryo 'regulates', or 'regenerates'.

Such phenomena are conventionally brought under the rubric of homeostasis, restoring the stationary equilibrium state, or *status quo*.

One could express this as a visual form in a diagram such as figure 3. Here the multidimensional spaces of genotypes and environments are represented as the plane $E \times G$, and the phenotypes corresponding to the genotype-environment coordinates within the circle drawn on $E \times G$ have values on a basin-like surface. Such a situation makes it impossible to use the conventional concepts of statistics, and to attach any precise meaning to Fisher's famous *Fundamental Theorem of Natural Selection*: 'The rate of increase in fitness of any organism at any time is equal to its genetic variance in fitness at that time' [3]. When genotype and fitness—which is a phenotypic character—are related in any way comparable to that diagrammed in figure 3, the concept of 'variance' simply cannot be applied.

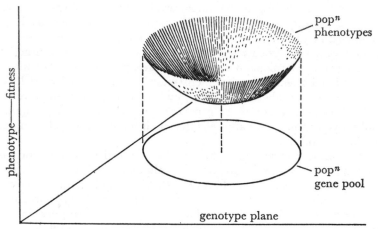

Figure 3. Relation of phenotypes to gene pool.

As an embryologist, whose basic interest is in the process of development, I can hardly be satisfied with such an atemporal concept; and I have urged that we should think not in terms of homeostasis, but rather of homeorhesis, the stabilization not of a stationary state, but of a pathway of change in time. Instead of picturing the resistance-to-change of a phenotype as a minimum in a space that is multidimensional for physicochemical variables but that does not include time, I envisage it as a set of branching valleys in a multidimensional space that includes a time dimension, along which the values extend. The development of the phenotype of an individual proceeds along a valley bottom; variations of genotype, or of epigenetic environment, may push the course of development away from the valley floor, up the neighbouring hillside, but there will be a tendency for the process to find its way back, not to the point from which it was displaced—the homeostatic equilibrium—but to some later position on the 'canalized pathway of change' (or 'chreod') from which it was diverted. Or, if it is pushed over a watershed ('catastrophe surface'), it may find itself running down to the bottom of another valley (figure 4).

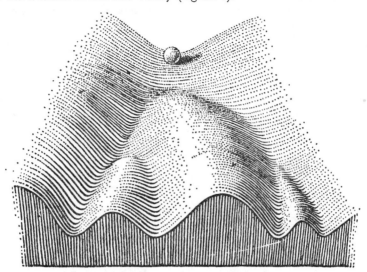

Figure 4. The epigenetic landscape.

This notion, of the canalization of the phenotype, is really only a return, with the benefit of more recent insights, to the old classical idea of 'the wild type'. In the days of Morgan and Bridges, in the twenties, it was remarked that all individuals of given species captured in the wild state (e.g. all individuals of *Drosophila melanogaster*) look remarkably alike; they were examples of 'the wild type'. Under the influence of the crusading zeal of those days to show that genes are at the centre of everything in biology, this observation was interpreted to mean that nearly all individuals have essentially the same wild-type genotype; only rare 'sports' carried, and showed the effects of, recent mutations of genes to abnormal alleles. Later work, especially that of Dobzhansky, showed that this interpretation is quite incorrect. The individuals that are phenotypically almost identical, looking as alike as two peas, contain wildly different genotypes, each a sample drawn from the population's highly heterogeneous gene pool. The uniformity of the wild-type is a *phenotypic* uniformity, a result of the canalization of development, which conceals the heterogeneity of the genotypes and of the epigenetic environments.

A mathematically more explicit exposition of the concept of an epigenetic landscape, in terms of a computer programme, is at present being worked out by Dr Adrian Walker of the State University of New York at Buffalo, N.Y. For my present purposes, however, I shall be content to set on one side the temporal dimension in the genotype-to-phenotype transition. The point I want to make can be expressed if we envisage the phenotype as a minimum on only two dimensions. We consider any particular phenotype P as located at a minimum in $E^\alpha \times G$, invariant against changes in either of those unless they surpass some stability limits.

We are dealing with a situation whose logical structure is that of the 'cusp' catastrophe (figure 5).

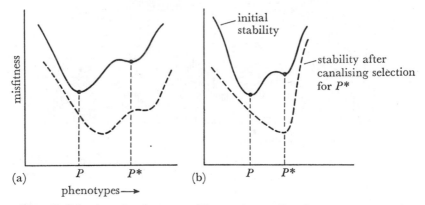

Figure 5. Selection of a phenotype. The continuous line shows the misfit of P, and the dashed line that of P^*.

On such a system, selection may operate in two different ways. The simpler will be selection for the actual size of P, which will have the effect of pulling P over to another phenotype P^* (figure 5(a)). A more complex form of selection, which I have spoken of as 'canalizing selection' [4], may act on the stability of characteristics of the system (dotted lines in figure 5(b)). The occurrence of this second type of selection has been observed in situations in which selection pressure continues for many generations.

The altered phenotype P^* will be at some sort of a minimum, and have some degree of stability, usually much less than that of the phenotype P which we suppose to have been subject to selection for many generations in a reasonable constant environment. If the abnormal environment is removed, P^* can easily slip back again to P. It may not be able to do this within the lifetime of a single individual, since phenotypes are normally much more flexible at early stages of life and soon become difficult or impossible to alter; but we must expect that the offspring of $P*$ will revert to P in the absence of the special environment. However, if selection operates to favour the transition from P to $P*$ under the influence of the special environment, new genotypes will be built up for the $P*$ individuals by canalizing selection, and these may, and in experiment are found to, modify the original stability properties in such a way that $P*$ becomes stable over a range of environments that include the original environment of P from which the special environment which initially produced $P*$ is absent (figure 5(b)). This is the phenomenon of 'genetic assimilation of an "acquired" character'. The prediction that it should occur [5] and the demonstration that it does so [6, 7] is the main experimental confirmation to date of the theoretical approach being sketched here; it seems doubtful if it could be reconciled with any theory that omits taking account of the stability properties of phenotypes, and therefore of the catastrophe character of phenotype alterations.

There is a further feature of phenotypic stability that is of great importance for the theory of evolution. When a change of environment is powerful enough to force a previously adapted phenotype out of its stable position into some new position, the character of the new phenotype is frequently such as to increase its fitness (or minimize its misfit) in the changed circumstance. This is a restatement of the well-known fact that organisms have a strong tendency to respond *adaptively* to novel demands made on them. I shall not discuss here the epigenetic mechanisms by which this result is brought about; they vary widely, from the appearance of such all-purpose mechanisms as the immunological system of vertebrates, to reactions that operate only over a much narrower range of environmental challenges. The first point I want to make here is to emphasize the philosophical importance of this adaptability. There has grown up, within the Neo-Darwinist paradigm of evolutionary theory, a dogma that the character of any new elements that may appear in a population as potential raw materials for evolution are quite unconnected with the nature of the selection process to which they will be subjected. A theory that attaches fitness coefficients directly to genotypes must see the raw materials of evolution as gene mutations, produced by a 'random' process having no connection with selective forces. This seems to be not only tenaciously believed, but deeply felt, as a reinforcement to a sense of the tragic, or tragicomic, existential situation of man. A recent widely acclaimed polemic on this theme has been Jacques Monod's book *Chance and Necessity*. According to the view advanced here, however, selection impinges on phenotypes, not on genotypes, and the gist of this paragraph is that the character of new phenotypes is, to some extent at least, influenced by environmental agencies (E^a) closely related to those (E^β) by which it will be selected; and, further, that the influence is often in a helpful direction. This is a conclusion not so well attuned to fashionable despair, but more in line with biological realities.

More appropriate to the present meeting is the question: how should one represent this situation—of adaptability—in mathematical terms? Here again I can offer only an amateur formulation, in the hope that it will provoke someone with more mathematical sophistication to propose something better.

We might simplify the specification of all possible environments to a two-dimensional plane, with the up-until-now standard environment E at a central point, and all changes from this to be indicated by a movement of the environment point in an appropriate radial direction. Then the phenotype adapted to the standard environment would be represented, in the third dimension, as a pit of minimum misfit, with the lowest point over E. The theorem of general adaptability would be the statement that, if the environment changes in any direction toward E^*, it will tend to push the phenotype into an already existing misfit-minimum that is waiting for it in this direction. Thus the misfit-minimum of P over E is surrounded by a ring of minima completely encircling E (figure 6).

environment plane

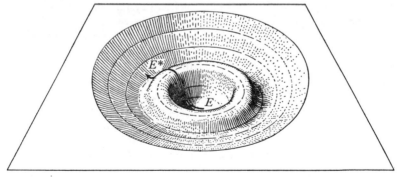

Figure 6. Misfit surface for phenotype exhibiting adaptation to changes in environment.

I think one would suspect that such a situation would be unstable, and that the continuous ring of E^* minima would break up into a number of somewhat separated minima. I confess that this is an intuitive rather than a rational suggestion. It would mean that some particular population of organisms would be better able to react adaptively to certain environmental changes than to others; they might be good at adapting to elevated temperature or humidity, not so good at dealing with reduction in supplies of a favourite food. Are there any symmetry considerations about the stability of such continuous minima-rings that issue from the mathematics, and that might be suggestive to the biologist?

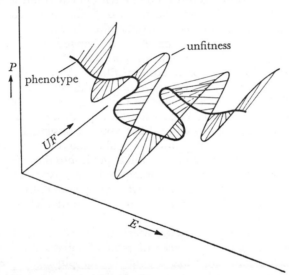

Figure 7. Three-dimensional diagram of the relation between environment phenotype and unfitness.

Actually, in figure 6, we have confounded the environment and the fitness. They can be separated only, in a three-dimensional diagram, if the environment is reduced to a one-dimensional space. Figure 7 is a three-dimensional diagram of a radial cross-section of the surface diagrammed in figure 6. Environmental variations are represented in one dimension from left to right; phenotype is also reduced to one dimension, vertical; and unfitness is in the third dimension, increasing away from the observer. The diagram illustrates the folds in the phenotype-misfit surface that are expressed as catastrophes when projected on to the (selective) environment plane. But the point I am making here, about the common tendency for environmentally induced phenotypic changes to be adaptive, is expressed by the lateral tongues of 'reduced misfit' protruding forward over each of the two minor regions of phenotypic stability on each side of the major one.

Acknowledgements

I should like to thank Professor Christopher Zeeman and Mr Maurice Dodson for helpful discussions.

DISCUSSION

DR SUDAK: I did not understand one of your figures. If you have a phenotype P^* which is unstable in environment E but is stable in environment E^*, then when you put it back into environment E, it becomes stable again. It sounds as if something else must have changed.

DR WADDINGTON: Something else has changed. In environment E^*, you have selected it. When it first goes into E^* it is a bit variable. In E^*, you selected it for maximum fitness in E^*; this changes, stabilizes, this phenotype and changes its stability properties.

It is a little difficult to explain on that diagram, which is a summary of it, but the whole of this has been published at length and discussed at length elsewhere. What you are essentially doing is pushing it out of one valley into a neighbouring one, and you may at first have to push it up a bit of a hill to get it into the neighbouring valley. If you then select it for going into the neighbouring valley easily, that hillside gradually cuts away, as it were, until there is not any threshold to push it over. It goes there naturally. That is another way of expressing it.

DR S.BAGHCHI (*Loyola College, Montreal, Canada*): You have been describing mathematical space, division of subgroups, and topological spaces, somewhat like kinematics for physics. Where does dynamics come from, because from one type you can get various types under different conditions, which in your case might be the environment. I would like to know how you can put dynamics in it. So far, from physics you are doing kinematics, if you go on mathematically from one group to the subspace, how can one put dynamics in it?

DR WADDINGTON: I suppose by 'dynamics' you mean the actual processes involved in changing the phenotype. I mean that these are embryological processes of various kinds, eventually syntheses of specific proteins, and what have you. They would be fantastically complicated. One of the points of mathematics, it seems, is that you can talk about the real problems without necessarily going into all the fiddling details. You could put in bits of dynamics, but you could not possibly put in all the dynamics of controlling a phenotype, partly because they are not known, and partly because they are much too complex.

DR JAMES SOLARMAN (*Hunter College, New York, N.Y.*): I am still not quite clear why you feel that the phenotype is buffered. I think there must be some examples where it is not quite as non-flexible as you make it appear.

DR WADDINGTON: I think it is because, experimentally, phenotypes are buffered. You can change many genes and nothing happens. You then say the genes are recessive, but it is only another way of saying that the phenotype is buffered. You can change the temperature, you can change all sorts of environmental conditions, and normally the animals come out looking much the same as before. There are some animals that don't and some that do. There are two major strategies in the evolution, of which the commonest is to buffer your phenotype and produce a good average animal. You bring the mouse up in a tropical greenhouse or bring it up in a cold store. In the tropical greenhouse it will have a fractionally longer tail and slightly thinner coat, but that is all. Otherwise, it is exactly the same as the one in the cold store.

A few animals don't buffer their phenotypes. The little crustacea in every little pond have slightly different shapes, and they are responsive to environmental changes. But, in general terms, most organisms are resistant to alteration. They regulate or regenerate or have dominant genes, or epistatic genes, or something, but they don't change without a pretty hefty push.

DR HOWARD LEVINE (*Columbia University, New York, N.Y.*): Dr Waddington, at the very end of his talk, made the point that we are dealing largely with non-linear phenomena, and he was incapable of dealing with non-linear phenomena, as are most mathematicians, and would we prepare some nice theory for him. I agree that it is possible, using a computer, to do things that are non-linear almost as easily as you do linear things, but then you come into another problem. I just wanted to emphasize that you have a great many variables, with all sorts of interactions between them, and you must simplify before you can make either a mathematical model or a computer model. I think the real problem is to decide what things we want to put in and what things we want to leave out, and once we have done that, we can get somewhere even if we don't know good mathematical treatment.

DR WADDINGTON: I think I quite agree with that. I don't think this was so much of a question as an amplification. Of course, we can make models only by leaving something out, and of course, the key question is, what do we leave out? This will presumably be different in relation to different questions we want to ask. I do not think I have any quarrel with what Dr Levine said.   [1974]

27

THE HUMAN ANIMAL

The biologist who looks from his professional standpoint at the human race sees man, of course, first as an animal: *Homo sapiens*, one of the species belonging to the family of primates, who are a subclass of the mammals and a branch of the great vertebrate stock. Even that bald identification carries with it many implications and it is as well to begin by enquiring just what they are.

From the earliest beginnings of scientific enquiry until quite recently, biology has been in two minds as to how to envisage the essential nature of animals and plants. One tendency has been to see them as nothing but rather elaborate machines. Descartes can be taken as an early and fairly extreme exponent of this view. The other tendency has been to suggest that, quite apart from any question of a specifically human soul in the theological sense, all animals and plants contain in their essence some non-material or vital principle. Even many of those who provided straightforward causal or mechanical explanations of some particular activities of living things have frequently argued that, over and above such detailed processes, or, if you like, behind them, there must be some essential, living, non-material agency. This was the view, for instance, of Harvey who, with his discovery of the circulation of the blood, actually did considerably more than Descartes himself to reveal some of the mechanical processes on which animal life depends. The logical opposition between these two views grew deeper as knowledge of material mechanisms became more clear-cut and more precisely formulated. It reached its height perhaps in the latter years of the nineteenth century, at a time when the physical scientists were profoundly convinced that matter consists of billiard-ball atoms and that is all there is to it. By this time the practical successes of physical theory were so great, and had won for it such a dominating position in scientific thought, that the few remaining vitalists, such as Driesch, had almost the position of isolated eccentrics.

Within a decade or two, around the turn of the century, the whole picture changed radically, and the long-standing 'vitalist-mechanist controversy' effectively vanished from the scene of biological thought. It disappeared because it was borne in on both sides that they had been over-simplifying matters. On the one hand, the physical scientists discovered that it is inadequate to reduce matter to a collection of impenetrable and unchanging billiard-ball-like atoms. They found themselves instead forced to think in terms of subatomic particles, wave-mechanisms, relativity and the interconvertibility of energy and matter, and even at a loss to support the principle of causal determinacy. No force was left in the statement that living things were nothing but matter, since it had transpired that matter itself was still a most incompletely comprehended mystery.

At the same time, thinkers about biology realized that when simple units become structurally arranged into complicated systems, these systems can exhibit new properties which can be understood by hindsight but not necessarily by foresight [1]. That is to say, certain properties of the units may never be exemplified except in the conditions created by the assemblage of the units into organized structural complexes. The crucial point is that one cannot expect, from examining the behaviour of the units in isolation, to deduce all the activities which may be shown by a suitably structured arrangement of them; any more than by looking at a few pieces of wire, glass, plastics, nuts and bolts, etc., we could deduce that when suitably arranged as an electronic computer they could beat us at chess.

It became obvious, in fact, that the explicative power of architecture or organization—what has sometimes been rather grandiloquently referred to as emergent evolution—is so enormous that any temptation to invoke a vitalistic principle over and above this, almost totally vanishes. We can safely say that living things are complex arrangements of 'matter', but since we have scarcely any clue to what matter is, and the main information we have about complex arrangement is that it is almost incredibly efficient at producing unexpected results, this statement can do little more than allay uncalled-for philosophical qualms, and in point of fact adds next to nothing to our understanding of the situation.

Biologists were then able to devote themselves with an open mind to the study of their proper subject-matter, the living world. One aspect of their endeavour has been to try to discover what should be taken as the basic units out of which living things are built. Putting it very briefly, the conclusion that has emerged so far is that the most characteristic processes of life depend on the activities of protein molecules operating as organic catalysts or enzymes which speed up certain reactions to rates much faster than they would otherwise show; but that the specific character of these enzymes is determined at a more fundamental level by the hereditary factors, or genes, in whose composition nucleic acids are more important than proteins, which an animal or plant inherits from its parents.

These studies on the basic mechanisms of living processes do not offer much illumination on the problems of how human life should be conducted. More suggestive insights arise in connection with the other major aspect of biological study, that is, the investigation of the ways in which the ultimate units are combined together. The most important point is an extremely general one, namely that all biological organization, whether of cells, individual organisms or populations, is involved in temporal change. Life is through and through a dynamic process. Any mode of thought which attempts to attribute to man or any other organism any form of unchanging essence, or any character that is conceived as *being* rather than *becoming*, flies in the face of our whole understanding of biology.

The flux of becoming which is so characteristic of all living things is perhaps most clearly and inescapably expressed in the phenomena of embryonic development. We can watch a fertilized egg begin its life as a small almost featureless lump of living material, and gradually develop into an adult of considerable obvious structural complexity. In many cases, for instance in birds, it carries out this performance inside an eggshell which effectively insulates it from outside influences, except of such a crude and general kind as a reasonable temperature. It is clear that the fertilized egg must already contain within it substances whose reactions with one another suffice to ensure the production of the various different organs and tissues out of which the adult is built.

One of the best analogies for the type of process that must be going on is the homely one of cheesemaking. A mass of milk-curds infected, perhaps by chance or by careful design, with appropriate strains of bacteria will, if left quite to itself in a cellar, pass in a stately manner through a series of changes by which it becomes metamorphosed into a Stilton in all its glory of ripeness. In a developing egg, the situation has many similarities with this, but is much more complex. In the first place, one and the same mixture can develop as it were into a Cheddar, a Camembert, a Brie, etc., as well as into a Stilton. The egg, composed of the cytoplasm together with a collection of hereditary genes, can develop into a liver as well as lungs, nerves as well as muscles, and in fact into a large range of sharply distinct types of cells and organs. It does not follow only a single pathway of change, but has a number of alternative possible pathways open to it, one part of the egg taking one path and another a different one. Again, it is a fact of observation that these pathways of change are rather resistant to modification. A part of an egg may develop into muscle or it may develop into nerve, but it is difficult to persuade it to develop into something intermediate between the two. Once it has started developing, for instance into muscle, it shows a strong tendency to produce a normal muscle even in the face of interferences that might be expected to divert it from its normal course and produce an abnormal end-result. The paths of change are, as I have said elsewhere, *canalized*. They are not like roads across Salisbury Plain, where it would be relatively easy to drive between them over the grass. They are more like Devonshire lanes; once you are in one, it is very difficult to get out again and you have to go on to where the lane ends [2].

These pathways of change, along which the various parts of the egg proceed as it develops, are inherent in the constituents of the egg at the time when it begins its development after fertilization. The specification of the direction the paths take, and the nature of the end-result to which they lead, is in the main carried out by the hereditary genes which the egg has received from the two parents. If one of these genes is changed, some of the paths will be altered and an abnormal end-result obtained. There is no simple English word which can be used for this concept of a pathway which is followed by a system, and whose characteristics are defined by the nature of the system which enters on it. I have suggested that we might call them 'chreods', from the Greek word χρή, necessity, and ὁδός, a path.

A system is exhibiting chreodic behaviour when it is changing along the course whose direction is defined by the system's own essential nature. It is not being chreodic in so far as it is diverted from this path by the accidents which it encounters on its way. One could, of course, discuss how far the development of individual human personality, or the socio-economic development of particular societies, are or are not chreodic in nature. Such questions are interesting but I do not think that our biological knowledge is necessarily very enlightening in connection with them.

It is more to the point here to turn to consider the other major type of temporal change with which biology is concerned. That is, of course, the process of evolution. The whole realm of living things as we know them today has been brought into being by evolution; and this, of course, includes man. The notion of evolution is by now not solely a theory about certain processes which may go on in the living world, but is one of the essential dimensions within which biological thought must take place. We cannot think of living things in modern biological terms without at the same time employing the concept of their evolution.

From the very beginnings of biological thought, for instance in the works of Aristotle, it has been clear to mankind that living things can be arranged in some sort of natural order; an order which in late medieval times was referred to as the Great Ladder of Being [3]. This stretched from the lowliest creatures, such as slugs and worms, through a series of intermediates to the lion, the lord of beasts, then to man, and then above him to the circles of angels and archangels. As this classification implies, untutored man has never hesitated to consider some of the classes of living things as lower and others as higher. Selfconscious and sophisticated thinkers may sometimes be heard to enquire by what right man classifies the living kingdom into a hierarchy in which—is it by chance?—he turns out to be at the top. Nearly all biologists, however, essentially agree with Aristotle in this matter, perhaps mainly for reasons rather similar to those by which Doctor Johnson refuted Berkeley; they would be willing to consider the claims of a worm to a higher status than man when the worm comes up and presents them. The overwhelmingly general view of biology, indeed, is that there not only is a natural order but that this is an evolutionary order, the higher stages having appeared on the earth's surface later than, and by derivation from, the earlier.

This type of evolutionary progression from lower to higher is technically known as *anagenesis* [4]. It has been discussed by many recent authors and in particular by Julian Huxley, who has emphasized the fact that it is by no means the only type of result that evolution brings about. As he points out, evolution may bring into being a type of creature which succeeds in surviving with comparatively little change through long periods of geological time, a process for which he uses the word *stasigenesis*. Again, another typical result of evolution in the non-human world is the breaking-up of a group of organisms by branching into a large number of species which differ in detail while still resembling one another in the broad outline of their type of organization—a process for which Rensch has coined another technical word, *cladogenesis* [4]. But these two kinds of evolutionary result are embroideries on a main theme; which is the succession, throughout the history of life on the earth, of a series of dominant types of organization, each a clear-cut advance on what went before—the unicellular organization succeeded by the multicellular, the primitive multicellular types, such as sponges, succeeded by more complex types such as sea-anemones and worms, those again by insects and fish, the fish by amphibia, reptiles, birds and mammals.

How, in terms of these concepts, do we see the situation of man? His appearance on the world scene is clearly not a case of mere stasigenesis, since he has changed from his non-human ancestors. Again, his mastery of conceptual thought and social communication mark off his biological organization as something radically different from that of his nearest biological relatives, the higher apes: he therefore cannot be considered the product of mere cladogenesis, but must be considered to have resulted from anagenesis, a real progressive change and not a mere modification in detail.

If one inspects the anagenetic changes which have gone on in the sub-human animal world, it is not too difficult to discern some of their general characteristics. For instance, one of the most important of them has been an increasing independence of the external environment, exemplified, for instance, in the evolution of creatures that can live on dry land or even in the air, as well as in the sea, and of animals which can maintain a constant body-temperature. Again, there has been an evolution of more precise and sensitive sense-organs, and a concentration of the nervous systems into a single central and ever further-evolving brain, leading to improved capacities of knowledge and feeling and awareness in general, and to the emergence of mind as an increasingly important factor in evolution. Both these trends can be considered as aspects of the evolution of an increasing capacity to make use of, or exploit, the openings for life offered by the earth's surface. Both also would lead to what, considered from the point of view of the individual, must be considered as an increased richness of experience. It is immediately obvious that the evolution of man is a further step in the same direction. No creature has been able to become so independent as he of the accidents of its environment; no creature has such facilities for experiencing not merely the elementary processes of the world, but the relations between them. The capacities with which man's evolution has endowed him are an immensely extended carrying-forward of the main progressive lines of pre-human evolution into radically new realms.

The most important respect in which the appearance of the human race extends the lines of advance of the sub-human world are in connection not with the results brought about by evolution, but with its very mechanism. Evolution depends, of course, on the passing from one generation to the next of something which will determine the character which that following generation will develop. In the sub-human world this transmission of what we may call, in a general sense, 'information' is carried out by the passing on of hereditary units or genes contained in the germ-cells. Evolutionary change involves the gradual modification of the store of genetically transmitted information. A few animals can pass on a meagre amount of information to their offspring by other methods: for instance, in mammals some virus-like agents which have effects very like hereditary factors may pass through the milk; in some birds the adults may serve as models whose song is imitated by the youngsters, and so on. Man, alone among animals, has developed this extra-genetic mode of transmission to a state where it rivals and indeed exceeds the genetic mode in importance. Man acquired the ability to fly not by any noteworthy change in the store of genes available to the species, but by the transmission of information through the cumulative mechanism of social teaching and learning. He has developed a sociogenetic or psychosocial [5] mechanism of evolution which overlies, and often overrides, the biological mechanism depending solely on genes. Man is not merely an animal which reasons and talks, and has therefore developed a rational mentality which other animals lack. His faculty for conceptual thinking and communication has provided him with what amounts to a completely new mechanism for the most fundamental biological process of all, that of evolution [6].

It is becoming common to say now that man must take charge in the future of his own evolution, but many who say this seem to be implying no more than that man must try to control the store of genes which are available and which will be available in later populations. In point of fact, the type of evolution of which man should take control is one which he has as it were invented for himself. His biological evolution—that is, the changes in the genes in future populations—will presumably continue, but these changes seem likely to be of relatively minor importance, at least in the near future, although they might eventually become a limiting factor [7]. For the alterations in which mankind is at present primarily interested—the types of change, let us say, which distinguish the societies which produced Newton, Shakespeare, Buddha, Confucius and Jesus Christ from scattered bands of neolithic hunters—the crucial evolutionary mechanism is one which depends on the socio-genetic transmission of information by teaching and learning.

If we can, in this way, see mankind, as at present the most

advanced phase in a process of progressive or anagenetic evolution in which the whole living kingdom is involved, it would seem to follow, clearly enough to convince most of those sympathetic to Humanist thought, that it is man's duty, not only to mankind but to the living world as a whole, to use his special faculties of reason and social organization to ensure that his own future evolution carries forward the same general trend [7]. This is, I think, the accepted Humanist position, as it is put forward for instance by Julian Huxley, Needham, and others, and accepted by bolder minds even among those who adhere to traditional religions, such as Canon Raven and Pierre Teilhard de Chardin [7]. I certainly do not dissent from the conclusions which such thinkers have drawn as to man's duty at the present time, but I feel that our actual understanding of the biological world and of man's nature allows us to carry the argument forward by two not unimportant steps. These arguments, which I shall now advance, are by no means yet generally accepted.

In the first place, we may ask whether the process of anagenesis which can be seen in the animal kingdom, and the farthest step in this direction which has been taken by the appearance of the human race, are mere contingent happenings, which have actually transpired but for which no underlying cause can be envisaged. I do not think so. I think one can see reasons why processes of an anagenetic kind must be among the types of change which evolution will bring about. The biological mechanism of evolution is, as we have said, founded on the genetic transmission of information from parent to offspring through the formation of gametes and their union to form fertilized eggs. This process, however, constitutes only the essential transmission by which the generations are connected. Several other components are necessary to make up the total machinery by which evolutionary change occurs. The best-known of these components is, of course, natural selection, which by favouring the reproduction of certain individuals more than that of others brings about alterations in the store of genes as they pass from generation to generation. But natural selection and heredity do not work alone. As I have argued in more detail elsewhere [6], we have to take account also of the capacity of animals to select, out of the range open to them, the particular environment in which they will pass their life, and thus to have an influence on the type of natural selective pressure to which they will be subjected. For instance, a rabbit or a blackbird, released among fields, will take refuge in the hedges or banks, while a hare or a lark will choose to live in the open grassland. And again, we should not forget the type of responsiveness which comes to characterize the various developmental pathways which the egg can follow, which has an influence on the effects which will be produced by any new hereditary modification that may occur. Thus, the complete evolutionary mechanism, or evolutionary system as I have called it, comprises at least four major sub-systems—the genetic system, the natural selective system, the exploitive system, and the developmental or epigenetic system.

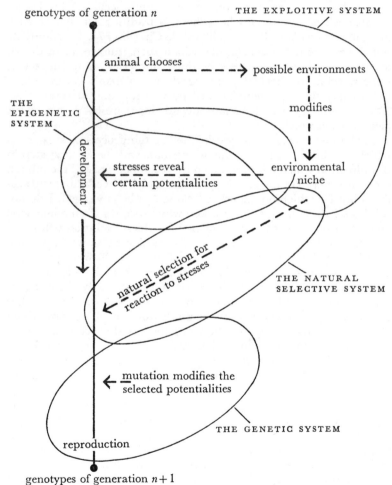

Figure 1. The logical structure of the biological evolutionary system.

Darlington [8] added a new dimension to evolutionary thought by pointing out that the genetic system would itself be subject to natural-selective pressures, and might itself evolve in such a way as to make it more efficient in passing on hereditary information in a form in which it is easily utilizable for the furtherance of evolutionary advance. For instance, the fully developed system of sexual reproduction found in the great majority of organisms, which is based on two sexes whose gametes unite to produce the offspring, is a very efficient mechanism for evolution, since it provides a way of recombining hereditary factors into a large number of new combinations, some of which may prove useful; but it itself is a considerable evolutionary achievement, since the most primitive living things, such as bacteria, do not possess it, though some of them have less advanced, so-called *parasexual* mechanisms which make some degree of recombination possible [9]. Now this same argument can be applied to the other sub-systems, and indeed to the evolutionary system as a whole. If we start with a world of living things capable of evolving, then not only will they do so, but the very pressures that bring about evolution will also tend to bring about an improvement in the mechanism by which evolution is mediated. Put in such abstract terms, this may sound a formidably complex notion, but actually it is easy to find quite everyday analogies for it. At the beginning of the Industrial Revolution, for instance, there were many factories capable of producing manufactured products; and the forces of competition between the factories, which we may for the purposes of this analogy compare to natural selection, not only brought about an evolution of the factory products (which correspond to the animals) into more elaborate and better fabricated articles, but equally brought to pass improvements in the organization of the factories themselves, that is to say, in the mechanisms by which the articles are produced. Again, to take another example, if a group of beginners take up the practice of playing card games with one another, they would not only become more skilful at playing the game they first start on, but are likely to pass on to playing subtler and more complicated games. Thus, this, as it were, two-tier evolution — an evolution of the end-product itself and also an evolution of the mechanism by which the end-product comes into being — is quite a normal sort of happening.

If we regard the biological evolutionary process from this point of view, we can see reasons why evolutionary changes, of the general character of those which are actually found, should have been expected to occur. One of the major components on which evolution depends is what we have called the exploitive system—the system by which animals choose and make use of the various possibilities for living which the world offers them. One of the evolutionary pressures which is bound to arise is, therefore, a tendency for an improvement in efficiency of the exploitive system. This is most clearly expressed in the evolution of the sense-organs and nervous system, and is, as we have seen, one of the major components of anagenetic evolution as we can trace it from the lowliest flatworms and jellyfish up to the higher vertebrates. Again, there will be evolutionary pressures acting to improve the genetic system. The enormous improvement—in rapidity of action, subtlety of recombination and regrouping of items, and so on—which has been brought about by the human sociogenetic system, as compared to the biological genetic system, can therefore be seen as one example of a general category of change which evolution must have tended to produce.

We can in this way at least begin to envisage the course of evolution as we find it, not as something completely accidental but an exemplification of general trends or types of change which we should expect. We shall perhaps never be able to assign precise reasons why that particular change which actually occurred was the one that did so out of all those possible. It is only in the broadest outline, when we are considering its general direction and categories of effects rather than particular effects, that we can see evolution as a chreodic process whose course follows from the characteristics of the system itself; but even an understanding in very broad outline is preferable to the state of complete incomprehension which can do no better than accept what it finds in the living world as mere 'happening to be so'.

Although we can see that there would be an evolutionary pressure towards the production of an improved system of transmitting information, and that if one were to appear which was in any way more effective than the biological genetic system, it would bring with it great evolutionary advantages, we still could not have foreseen that this step would have been taken by means of the very remarkable and peculiar mechanism which seems to characterize the human species. Even the remarkable work which is now being carried out on the behaviour of sub-human animals, in which the sociogenetic stage has not yet been attained, gives us little hint of what to expect[10]. Just how remarkable the human system is has only recently been brought home to us, largely as the result of the work of the psychologists.

It is clear on first principles that any system of social transmission of information can only operate if in some way the potential recipients can be brought into a condition when they are ready to accept the content of the messages which are directed at them. In man, it appears that the moulding of the newborn infant into an effective recipient of social communications involves a most surprising process of projection and re-introjection of certain of his own impulses, together with the building up of internal representatives of parental authority, and a whole peculiar mechanism which is described in terms of such concepts as the super-ego, the ego-ideal and so on. At first sight, the story the psychoanalysts tell may seem extremely unlikely, but it seems to me they have now produced enough evidence to render it rather plausible, at least in broad outline; and on reflection one realizes that unless one is prepared to make the question-begging assumption that man is simply born socially receptive, some sort of process or other would have to be imagined by which he is brought into this condition.

Now, the second point I wish to urge, in extension of the normal Humanist argument, is that man's ethical feelings are essentially involved with, and in fact are actually a part of, the mental mechanism by which he is developed into a being capable of receiving and accepting socially-transmitted information. Unless some sort of authority-bearing system is developed in the mind of the growing individual, social transmission would break down because nobody would believe what they were told. One part of this authority-bearing system develops into what we call our ethical beliefs, to which indeed we usually attach an almost overwhelming authoritativeness. Another aspect of the system seems to be, unfortunately, a tendency to develop feelings of inferiority, guilt and anxiety — a situation in which one may, perhaps, glimpse from, the scientific angle of approach, the human predicament which is enshrined in the myth of the Fall of Man [11].

Obviously more than mere acceptance of authority is involved in a fully developed system for the social transmission of information. One can, and in later life one must, compare what one is told with objective reality, and reject what proves false. Education is to some degree concerned with such corrective verification. But all this is really a second-order process. There must first be a reliable system of transmission, which corresponds to biological heredity, before there can be a process of verification, which we might compare to natural selection. Again, it is certainly true that man's innate genetic constitution provides him with potentialities, which are presumably absent or very weak in other animals, for developing his social transmission mechanism. One of the most impressive pieces of evidence for this genetic predisposition is provided by the life of Helen Keller, who although blind and deaf from early infancy, nevertheless came to grasp the fact that 'things have names', and thus showed that she had the basic faculty for apprehending language [12]. But it is only with the development, normally in the first few months of life, of these innate capacities to the point where the child accepts transmitted information that man's secondary evolutionary system begins to function.

If this argument is accepted, the connection between evolution and man's ethical nature is much closer and more intimate than even most Humanists have previously recognized. It is not merely the case that we can see ourselves as part of an all-embracing process of evolution and therefore can recognize a duty to further the general evolutionary tendencies. According to the argument advanced above, man is characterized by the emergence of a new evolutionary mechanism based on sociogenetic transmission, and in this transmission the development of something akin to ethical belief is an absolutely essential item in the mechanism. The orthodox Humanist argument is that it would be a recognizably good thing if we took steps to see that our ethical beliefs effectively controlled the further course of evolution. What I am arguing is that our ethical beliefs *must* influence the course of human evolution, since that is based on a mechanism of which those beliefs are an essential part. The question that is really at issue is not whether evolution shall be guided by ethical beliefs, but what kind of ethical beliefs shall guide it.

What the situation of man calls for, in fact, is the formulation of some criterion by which one could judge as between the various ethical beliefs to be found in different individual men and women or different human societies. It is not sufficient that Humanists should demand that future human evolution should be guided by ethical principles, since inevitably some sort of ethical principles—quite possibly, as the psychoanalysts have taught us, unconscious or only partially conscious ones—will in fact play an essential role in bringing it about. What we should be aiming at is that the ethical principles themselves should be subject to assessment according to some more inclusive criterion. The real contribution of the study of human biology and human evolution will come when it is used to help in the formulation of this supra-ethical criterion.

If the essential reason why mankind develops ethical beliefs at all is because this is necessary as an essential cog in the machine of social transmission by which human evolution is brought about, then it follows that we can judge between different ethical systems by considering how far they fulfil their function in furthering human evolutionary progress. I am not for a moment suggesting that we shall find it easy to reach a clear, let alone an agreed answer, but we shall at least know what we are trying to do, and this, though by no means easy, is well worth doing—for instance, when one is weighing against one another the values of individualism and collective organization, of nationalism and internationalism, of increase in population numbers and increases in standard of living, and so on through the list of the major moral and social quandaries of today.

The basic Humanist position, derived from considering man's place in the biological world, is that in approaching such problems we have to consider them in relation to what we know of the actual *course* of progressive evolution in the sub-human, and in particular of the human world. The arguments I have put forward in the last few paragraphs, although they go beyond the orthodox Humanist case, only serve to reinforce its conclusions. Evolution is the very essence of living. Life could, indeed, be defined as the state of a system which is capable of evolving, and the essential characteristic of man—if you like to put it so, the 'soul'—which distinguishes him from the animals, is that he evolved by a mechanism that belongs to him alone, and which he alone can modify and improve.   [1961]

POSTSCRIPT [13]

An extract from a letter which C. H. Waddington wrote to Sir Alister Hardy, putting his views on Organic Selection and genetic assimilation, in connection with comments which Sir Alister had made in his Gifford Lectures [14].

'The part of the book I have so far looked at most thoroughly is, as you might expect, your discussion of the Lloyd Morgan and Baldwin ideas about Organic Selection. In spite of what you say I still think there is something rather different about what I said from what they said. Please don't think I say this because I want to ensure any sort of 'priority' for myself. I quite admit that my ideas are extremely close to theirs. They were however not actually derived from them. I don't believe I had ever heard of Baldwin and Lloyd Morgan when I first thought of the idea of genetic assimilation around 1942.

'The real difference that there is, in my opinion, between our ideas arises in fact just because we approach the same basic notions from different directions. You say on page 171 that between Darwinism and what you are talking about there is a 'real if somewhat subtle difference: a difference which it is essential that we should understand . . .'. I entirely agree with you and I also agree with your earlier remark (p. 163) that 'this is not just a slight subsidiary effect

but is indeed one of the major factors in the evolutionary process.'
But I think the point I want to make adds on to yours, in just as
important a way as yours adds on to Darwinism.

'If I may somewhat exaggerate the matter to bring it into clearer
focus, I should put the situation as I see it like this. The early writers
about Organic Selection saw the importance of behaviour or habit
and this is the point that appeals to you as a naturalist. But I think
they did not completely see the point that the entities which undergo
evolution are not simply populations of genotypes but are populations
of developing systems; that is to say, organisms one of whose essential
features is to undergo development, and moreover, development in
which the environment plays a role as well as the genotype. In my
opinion the conventional Neo-Darwinist theories of Haldane and
Fisher (and to a lesser extent, Sewall Wright) are inadequate *both*
because they leave out the importance of behaviour in influencing
the nature of selective forces, *and* because they attach coefficients of
selective value directly to genes, whereas really they belong primarily
to phenotypes and only secondarily to genes. I still doubt whether
Lloyd Morgan and Baldwin had got the second point.

'The way this emerges in their theory is as follows. Lloyd Morgan
supposes that a population of a "plastic" species can survive under
new conditions of environment (p. 167, point 8 [15]). He even sees
that there will be some genetic variation in plasticity (point 9) but
he explicitly separates the environmental modification from anything to do with the genotype (point 10). In points 11, 12 and 13 he
seems to me to be postulating the occurrence of new gene mutations
in genes quite unconnected with those involved in the variation of
plasticity, these mutations being of a kind which tend to cause the
organism to develop the appropriate modifications independently of
any action of the environment. At least I have always thought that
the point made under 14 on page 167 meant that there was a con-
genital predisposition to develop to the modified phenotype without
any contribution from the environmental circumstances. This is also
the way that Huxley (cf. your p. 163) and Simpson have interpreted
Lloyd Morgan. I suppose however that it would be possible to interpret it in a sense much closer to my genetic assimilation. The concept
I want to resist is that the organism contains one set of genes which
allow it to adapt to the environment and another set of genes in
which mutations produce the adaptively modified phenotype
independently of environmental influences.

'As to the importance of behaviour in evolution, I think I have
been moving closer to your position and realizing more fully its

central importance. I already began to have ideas along these lines quite a long time ago. In fact I wrote a paper about "Environment selection by *Drosophila* mutants" in *Evolution* in 1954 [16]; I made a point that I notice you lay some stress on, namely that cryptically coloured forms have to behave appropriately *if* they are going to derive any benefit from their potential camouflage. At that time I wrote or spoke to Kettlewell and asked him if his light and dark moths did actually settle on the appropriate parts of the tree trunks, and a year or two later he published evidence that they did so [17]. In recent writings I have been stressing such points more and more, and suggesting that the next step in the mathematical exploration of evolutionary theory will be the application of Games Theory.'

28

THE HUMAN EVOLUTIONARY SYSTEM

Very soon after the publication of Darwin's work attempts were made to show how the development of human societies and civilizations could be explained in terms of processes similar to those which Darwin had shown to be operative in the sub-human world. Unfortunately, the discussion of the early social Darwinists did not prove very enlightening. Their influence was perhaps strongest in America. The school which flourished there around the turn of the century has recently been critically evaluated by Hofstadter [1]. The inadequacy of their account of social processes arose essentially from an attempt to apply to human societies a rather crude interpretation of the hypotheses which Darwin had advanced concerning the animal world. In particular an unhappy role was played by that most misleading slogan, 'the survival of the fittest'. It seemed to follow from this that the doctrine of natural selection implied that those who were most successful in society were necessarily more important for the furtherance of evolutionary progress than their fellows. The multiple imprecision which the phrase actually contains was only gradually realized.

THE CRITERION OF NATURAL SELECTION:
HEREDITARY TRANSMISSION

We are all fully aware now that natural selection is not primarily concerned with survival, in the sense of the persistence of any single lifetime, or indeed in any sense in which it can be validly equated with success. Evolution is a matter not of single life-times but of the passage of generations. What is important for it is not survival but transmission of qualities to offspring. This is, indeed, of absolutely overriding importance; so much so that when we speak of those which survive, or, better, transmit, as being the fittest, we are really adding nothing to the statement that they transmit. Fitness in this context must be defined in terms of transmission. The doctrine of the survival of the fittest becomes translated on analysis into the tautological statement that those individuals in a population which are most apt to leave offspring ('fittest') will leave most offspring ('survive').

With the breakdown of the early attempts to apply a crudely-interpreted Darwinism to social affairs, the two studies, the evolutionary biology of the subhuman world and the investigation of human societies, have tended to pursue different paths. Of course, sociology, like all other departments of human thought, has been profoundly affected by the general philosophical implications of the theory of evolution — the substitution of dynamic for static categories of thought, the interest in origin and becoming rather than in mere being; but the more technical biological theories, and the developments which they have undergone more recently in the hands of students of the non-human world, have been of comparatively little importance to sociology. The time has now perhaps come when it may be worthwhile to reopen the discussion of the relevance of our present understanding of biological evolution to the processes of human social change. It is not to be expected, as we shall see, that biology can provide answers for any of the problems of sociology; but it may be that the modes of analysis and the types of formulation which have been found useful in the study of animal evolution will suggest questions which it may be profitable for the sociologists to ponder in relation to his own subject-matter.

ADVANCES IN THEORY SINCE THE TIME OF DARWIN

The theory of biological evolution has made several advances of a fundamental character since Darwin's day. Perhaps the two most important, and the two which occurred earliest, were the more or less simultaneous realization among biologists that the Lamarckian hypothesis of the inheritance of acquired characters could not be accepted as a general mechanism, and the rise of Mendelian genetics. In Darwin's day, biology possessed no proper theory of heredity. It was thought, in a rather vague way, that when two parents had offspring, their hereditary qualities came together and formed a blend in their progeny. Darwin realized that this would entail a rapid loss of individual variation, which would tend to disappear in a general uniformity. If this were to happen, there would be no raw material on which natural selection could operate, and the Darwinian mechanism would be rendered inoperative. According to the ideas of heredity current in Darwin's day, evolution demanded the operation of some system which engendered new hereditary variation as fast as cross-breeding led to its disappearance. Darwin, though with considerable hesitation, was tempted to find this source of variation in the effects of the environment, as Lamarck had suggested. But the accumulation of experimental evidence gradually made it seem less and less probable that characters acquired by interaction with the environment are inherited, until about twenty years ago this theory was almost uniformly rejected as incorrect. Since then in quite recent years a few scattered examples of phenomena have been discovered which could be interpreted in this sense. It is, however, still clear that, although in certain circumstances a character acquired by an organism during its lifetime may be passed on to its offspring, such a transmission occurs only in exceptional and particular circumstances. It is highly improbable that processes of this nature can play any large or important part in the main course of evolution.

MECHANISMS OF HEREDITARY TRANSMISSION

As this hypothesis gradually lost its attractiveness, the new discoveries of genetics revealed for the first time the actual mechanisms of hereditary transmission in the biological world. The demonstration that heredity is controlled by genes, which do not blend with one another but retain their own character except when they occasionally alter by the process of mutation, entirely removed Darwin's difficulty about the disappearance of hereditary variation in cross-breeding, and rendered unnecessary any reliance on Lamarckian theory as a source of new variation. It is on the facts of genetics and the conceptual framework which has been elaborated to account for them that the whole of our present understanding of biological evolution rests. We shall shortly have to discuss them in some detail. But before doing so it is necessary to point out that the history of evolutionary theory did not cease with the rediscovery and elaboration of Mendelian genetics. We have now reached a stage at which the discoveries of two other branches of biology are beginning to play a very important role, which will almost certainly increase markedly in the near future. These two sciences are the study of development, also known as experimental embryology, or, as I prefer to call it, epigenetics, and, secondly, the study of animal behaviour, sometimes known as ethology.

MEMBERS OF THE EVOLUTIONARY SYSTEM

The whole set of processes and mechanisms by which evolutionary change is brought about may be called 'the evolutionary system'. If we take into account the contributions of all the various branches of biology mentioned above, we have to consider the evolutionary system as composed of four main sub-systems. One of these is the 'genetic system', that is to say, the biological mechanism by which hereditary variation is brought into being and transmitted from one generation to the next. Another sub-system may be called the 'epigenetic system', and comprizes the mechanisms by which the specification (or information) contained in the fertilized egg is translated into the form of the adult organism capable of reproduction. A third factor is the 'exploitive system', by which organisms may select out of the range of possibilities open to them the particular habitat in which they will pass their lives, and which indeed they will often succeed in modifying. Finally, we have the natural selective system, i.e. the pressures and stresses of various kinds originating in the environment, including the biological environment, which determine the success of the various individuals in achieving reproduction. These four sub-systems are, as we shall see, not separate from one another, impinging individually on the evolving animal, but are instead intimately interdependent, so that the whole complex forms a cybernetic system involving feed-back relations.

Before starting our consideration of human evolution it will be as well to say rather more about the nature of the biological process, and in particular to try to illustrate in more concrete terms what is meant by the use of such grandiloquent terms as cybernetics in this connection. The genetic and the natural selective systems are the two components of the total evolutionary system which have been most fully discussed. It will be necessary to consider the genetic system in some detail when dealing with its possible application to the mechanism of human evolution, but there are no new points of principle that need be brought out at this stage in the argument.

THE EXPLOITIVE SYSTEM

The exploitive system has not in recent years been so much in the centre of the stage of biological interest, but its operations are relatively easy to understand. Naturalists have described many examples in which local races, or recognizable varieties within a species, show a tendency to choose a particular habitat, or exhibit a specific type of behaviour which brings them in contact with some corresponding aspect of the environment. For instance, in the butterfly *Colias eurytheme* in California there is a white form which is particularly active in the early morning and late evening, whereas the more usual yellowish types do not show a preference for these times of day [2]. Investigations of laboratory animals have also shown that different strains may be characterized by behaviour traits which cause them to choose different habitats. For instance, a large population of *Droso-phila*, consisting of a mixture of strains each of which could be recognized by the presence of some visible gene mutation, was released in a space out of which there were eight openings leading into compartments with all combinations of light against darkness, heat against cold, and wetness against dryness. The percentages of the different strains which moved into the various compartments were markedly different. The strains differed only not in their tendency to move at all (i.e. general activity) but some showed a specific tendency to go into certain types of environment rather than others [3]. Little attempt has yet been made to show conclusively that such preferences are inherited, but the fact that they are repeatable characteristics of particular strains indicates very strongly that they must have an important genetic component. There can be little doubt that if selection was exercised for or against a particular behaviour preference this trait would either be strengthened or weakened.

Another type of behaviour preference which may have important evolutionary consequence is that shown in the selection of sexual mates. Here again, naturalists and students of wild population have shown clearly that local races of animals may exhibit marked mating preferences, usually favouring mates of their own race. Similar differential mating preferences may be discovered in laboratory strains of a single species. In this case selection experiments have been made, and it has been possible to strengthen the originally weak tendencies within natural populations [4] and in laboratory strains [5].

It is clear that such genetically-determined behaviour tendencies towards preferences for particular types of habitat or mate will strongly influence the kind of natural selective pressures to which the individual is subjected, and thus the operations of the genetic system by which his inheritable qualities are passed on to the next generation. This provides a typical 'feed-back loop'. The nature of the genes in the animal helps to specify (or provides information for) the nature of the environment, and the environment, by its natural selective action, then helps to specify the character of the genes which will be contained in the next generation.

THE EPIGENETIC SYSTEM

The manner in which the epigenetic system operates in evolution is perhaps not so obvious at first sight. There are two important features of the developmental processes of animals which are particularly important in this respect. The sequence of changes by which the fertilized egg becomes an adult animal always involves considerable interactions between neighbouring parts of the embryo; for instance, mesodermal structures may induce specific types of differentiation in the ectoderm with which they are in contact, the tensions of the developing muscles have an influence on the form of the bones to which they are attached and so on. These mutual influences are often reciprocal and are highly complex. The result of their action is that the various organs and sub-systems which come into being have been to some extent moulded by each other, so that they tend to form relatively integrated systems, whose unity we frequently acknowledge by using the name 'organ'. The second feature of epigenetic processes of particular importance to evolution is the observation that they normally exhibit a subtle balance between flexibility and the lack of it. The flexibility is shown by the fact that an environmental influence impinging on the developing organism will often succeed in causing it to diverge, to some extent, from its normal course, and develop into an environmental modification; the organism develops an acquired character. But equally obvious is a certain lack of flexibility; development has a strong tendency to proceed to some definite end point. For instance, the adult tissues such as muscle, nerve, lung, kidney, etc., are quite distinct from one another and it is rather difficult to persuade developing cells to differentiate into something intermediate between these main types. Again, the animal as a whole will very often succeed in 'regulating', that is to say, in reaching its normal adult state in spite of injuries or abnormal circumstances it may have met during the course of its development.

Both these features, the tendency towards organization and the balance between flexibility and the lack of it, are characteristics of the epigenetic processes of comparatively highly evolved forms such as the multicellular animals. The study of the simplest living systems has shown very clearly that the ultimate units of the developmental processes are syntheses of single specific chemical substances, carried out under the control of nuclear genes. One can detect these in higher organisms also, but in them the entities with which the evolutionist normally deals, such as the limbs, digestive system, hair and so on, are complex systems involving the interaction of very many unitary processes of synthesis. It is to the nature of the interactions between these unitary processes that one must look for an ultimate explanation both of the organized character of the resulting organs and of their tendency to reach one particular end-point. The system of gene-controlled processes seems to be such that it has a certain number of relatively distinct courses of change open to it. We may say that the fertilized egg comprises a set of genes and associated cytoplasmic material which gives it the potentialities for proceeding along one or other of a definite set of trajectories of change as time goes on. Such pre-determined time-trajectories of change I have proposed calling 'chreods'. Each chreod must be thought of as having certain characteristics of stability (or rather of a quality similar to stability but involving time, which I have spoken of as homeorhesis); that is to say, the developing system will tend to move along a chreod and will resist being pushed away from its normal course by influences impinging on it from outside. If these succeed in diverting it for a time it will tend to come back on to the chreod when the disturbing influence is removed, though, of course, it may not always succeed in doing so completely, and then the organism, even when it becomes adult, will exhibit some modification or abnormality.

The relevance of this to evolution arises, in the first place, because there is likely to be some genetic variation in the stability-characteristics of the epigenetic systems of the different individuals in a population. Putting it in simpler terms, if a normal heterogeneous population is submitted to some environmental stress certain individuals are more likely than others to be affected and 'acquire a character'. Moreover, some will probably acquire characters which are more useful in furthering reproduction than those acquired by others. We shall have natural selection operating to increase the frequency of genotypes which enable their possessor to become adaptively modified to this stress. Moreover, the organizing capacity of epigenetic processes makes it likely that the acquired character will be a relatively harmonious one. It will not be a matter of, for instance, getting large muscles but lacking large bones to go with them. Finally, we have to take account also of the resistance to change shown by epigenetic systems. If selection for increasing the efficiency at acquiring a character has gone on long enough, it may make the developmental system so ready to produce the modification in question that it continues to do so even after the environmental stress is removed. When that happens we can say that the genetic system has 'assimilated' the environmental character. The total result exactly mimics the effects which might have been attributed to a direct inheritance of acquired characters, although the mechanism by which this has been brought about depends on strictly orthodox processes, and is quite different from that usually envisaged by those who still believe in the Lamarckian hypothesis.

INTERVENTION OF NATURAL SELECTION

These points have been expressed in a theoretical manner in order to give them generality, but they do not rest on purely theoretical reasoning. The fact that populations often contain genetic variation in the ability of the individuals to acquire characters; that they respond to selection for efficiency in this respect, and to selection for one particular type of developmental modification rather than another; and finally that such selection may lead to the point of genetic assimilation of the acquired character, have all been demonstrated experimentally [6]. The experiments also provided an actual example of a further consequence that one would expect to arise. As the genotype of an evolving population becomes moulded by selection for its capacity to react in a satisfactory manner with environmental stresses, it will have built into it certain chreods with particular homeorhetic characteristics. When new gene mutations now occur, by what we may consider random changes in the nucleo-proteins of the chromosomes, the effects these mutations produce on the phenotype of the animals containing them will not be entirely random, but will be, to some extent at least, characteristic of the epigenetic system in which the gene is operating. We have here another example of a feed-back loop, one which ensures that the phenotypic alterations produced by mutation are not completely independent of the demands which natural selection is making on the evolving animal.

HUMAN EVOLUTION, CULTURAL AS WELL AS BIOLOGICAL

It is now time to consider human evolution. The four-membered evolutionary system which has just been described must, of course, continue to operate in the human species, as in every other. However, the salient feature about man—perhaps one might say that it is his defining characteristic—is that he has developed, to an enormously higher degree than is found in any other species, a method of passing information from one generation to the next which is alternative to the biological mechanism depending on genes. This human information-transmitting system is, of course, the process of social learning. This gives man a second evolutionary system superimposed on top of the biological one, and functioning by means of a different system of information transmission. Most of the human evolutionary changes which seem of real importance to us—most, for instance, of the features which distinguish modern man from his ancestors of the Old Stone Age—seem to have been in the first instance produced by the action of the second evolutionary system dependent on social transmission. On theoretical grounds, one may be fairly confident that biological changes have been occurring at the same time, but it seems difficult to argue that they have been the primary variables in the process, and indeed it is not at all clear that they have any important relevance. It is, of course, an important task for human biology to attempt to elucidate the nature and magnitude of the genetic changes involved in human evolution. But this is only a part, and in my belief a subsidiary part, of the attempt to apply Darwinian or evolutionary thought to human affairs. The major task which confronts us now is to investigate the parallels and contrasts between the specifically human evolutionary system based on social transmission and what we know of the biological evolutionary system based on genetics.

Let us consider, then, the various factors which make up the biological evolutionary system, and see what parallels we can find for them in the specifically human evolutionary system. In biological evolution transmission of information from one generation to the next is in the main carried out by the genetic system. In this the transmitting event occurs at the time of fertilization, when each parent contributes to the newly formed individual a set of hereditary units, or genes. These units are essentially quite separate from one another, but in practice they are usually associated together in groups, and that in two different ways. As far as the biological genetic system is concerned, the most relevant form of association is a quite contingent one, which depends on the fact that the genes are located on chromosomes. Their arrangement into groups of neighbours held together by the material structure of the chromosome is usually, though not always, more or less irrelevant to their functioning. Another type of association between genes arises during development. The formation of the different tissues of the body, such as muscle, nerve, etc., depends on the inter-related functioning of sets of genes. In a certain sense, therefore, each specialized tissue of the body represents the activities of a group of genes (or perhaps better, represents a group of gene activities), the association of the members of the group being in this case not at all a matter of choice, since it is essentially dependent on the way in which the gene activities interlock and interact with one another.

These epigenetic groupings of genes—that is to say, groupings which arise from their activities during development—are not reflected in the normal transmission of information between generations of animals by the genetic mechanism. It is relevant to the human situation, however, to remember that even in animals something more may be involved between generations than the standard genetic system concerned with nuclear genes. The mother always contributes to the new animal a relatively massive amount of egg cytoplasm. In some groups of animals, for instance in insects, during the maturation of the egg in the maternal ovary a large amount of the contents of certain other cells, known as nurse cells, is bodily injected into the developing egg. These nurse cells could be taken to represent one particular epigenetic grouping of gene activities, which thus becomes to some extent transmitted to the next generation. Again, in many animals which bear their young alive there are formed, by developmental processes in the mother, mechanisms for passing into the offspring certain of the results of the activities of epigenetic gene-groupings in her body. For instance, substances from the maternal blood may be passed into the offspring through a placenta or similar structure. The most striking of such developments in the animal world is, of course, the development of milk secretion in mammals. We know several clear-cut examples in which transmission through such 'para-genetic mechanisms', as they might be called, can be shown to produce easily demonstrable effects on the offspring. Typical examples are the milk factor concerned with tumour development in mice, and the effects of the maternal body-size on the growth of the young embryo in reciprocal egg-transplantations between mammals of different sizes. In the subhuman world, however, the para-genetic mechanisms which can be utilized for transmitting to the next generation the results of the epigenetic interactions between genes are only rather slightly developed.

The situation is very different in the specifically human evolutionary system. The cultural, or as we may call it, the 'sociogenetic', transmission mechanism does not operate entirely or even mainly at one point in the life history of the new generation. There is no single entity that has to carry the message from one generation to the next as the gametes do in the biological world. We have, as it were, an enormous expansion and multiplication of modes of para-genetic transmission. An individual can receive information from his forebears throughout the whole, or at least the greater part of life, although, as we know only too well, the task of getting new ideas into his skull becomes progressively harder after a certain age.

This escape from the domination of a single major transmitter such as the gamete makes it possible for the sociogenetic system to handle groups of units which are associated, not only by chance as are the linkage groups of genes on the chromosomes, but which are grouped together by their functional interactions, in a manner comparable to the organized groups of gene activities which arise during development. If one were to attempt, therefore, to break down the content of social transmission into a series of unit items one should expect to find these items associated into groups in two rather different ways. Firstly, there are, of course, many functional interrelations between different items of transmitted content; the varied elements involved in a complex industrial technique, or the interrelated beliefs comprising one particular church doctrine, would provide examples. It is clear that the sociogenetic mechanism is quite capable of handing on such organized groups of units, the organization being a functional one and comparable to that which characterizes the tissues of an animal's body; that is, it is of a kind which in general is not capable of biological transmission to the next generation. But we might ask if there are also chance associations which could be regarded as comparable to the linkage between genes which happen to lie on the same chromosome. For instance, in the transmission of western culture to oriental nations it is very common to find that an item such as the wearing of western dress tends to be associated with other items such as various industrial techniques, or belief in the dogmas of Christianity. There is clearly very little essential functional connection between industrialism and the wearing of trousers rather than a sarong, dhoti or kimono. The association between such items, in so far as it exists, would seem to be a purely contingent one, quite comparable to that between different genes in a single linkage group. Social anthropologists seem in recent years to have been so interested in establishing the reality of the functional connections between elements of culture that the possible importance of purely fortuitous associations, comparable to those of linked genes, has perhaps been somewhat neglected.

GENETIC TRANSMISSION AND LEARNING

As a matter of fact subhuman evolution has involved not only the true genetic mechanism, and the para-genetic mechanisms such as transmissions through mammary secretions. There is a further subtlety, which is concerned with questions which almost merit the name of 'metagenetics'. As Darlington [7] in particular has emphasized, any particular mode of genetic transmission, for instance one depending on chromosomes and bisexual reproductions with cross fertilization, endows the organisms which utilize it with certain capacities for evolutionary modification. Another mode of genetic transmission, for instance one in which the organisms are self-fertilizing hermaphrodites, carries with it different capabilities for performing evolutionary advance. During the existence of living things on earth there have been, Darlington claims, alterations in these modes of genetic operation; there has been an 'evolution of genetic systems', as he terms it. The most important steps in this have been the organization of genes into chromosomes, the adoption of bisexual reproduction, and of course the evolution of the socio-genetic mechanism itself. But within the sociogenetic mechanism, once it has appeared, we can find a parallel process, an 'evolution of sociogenetic mechanisms'. In particular, there has been a develop-ment of phenomena of the kind which Bateson has called deutero-learning, that is to say, learning to learn. In this, the content acquired from the learning process is the ability to learn others things more quickly or more efficiently. This is an example of a second-order improvement in the mechanism of social transmission, comparable to the improvements (from the point of view of evolution) in the genetic system discussed by Darlington.

The development of deutero-learning introduces concepts which we might consider 'meta-sociogenetical'. There have, however, also been many evolutionary advances of a less radical nature in the sociogenetic system. They have produced alterations, which from the point of view of human evolution can in the main be considered as improvements, in the mechanisms of social transmission. Some of them have recently been discussed by Mead[8]. There is, for in-stance, a very primitive kind of social transmission of experience, which in fact many animals other than man also exhibit. In this, transmission occurs direct from individual to individual, through unverbalized behaviour of the teacher which conveys some inarticu-late message to the learner, either as a model to be imitated or as a directive to some course of action. This primitive mode of trans-mission often persists as an accompaniment to the more highly evolved sociogenetic mechanisms. Posture, gesture, turn of phrase, stress and accent all convey whatever it is they do convey in this manner. Although it is primitive and undifferentiated, this mode of transmission is not without power. The characteristics which dif-ferentiate the products of the best public schools and Oxbridge from those of the secondary modern and Redbrick have been in the main passed over by this mechanism which man shares with such other social creatures as the red deer and the prairie dog.

The main defect of this model-mimic or leader-follower system of transmission is its relative inefficiency in handling items of informa-tion which are capable of being conceptualized. It can indeed trans-mit emotional or affective material which is by no means simple, but there is a great deal of human experience for which other systems of sociogenetic transmission have, it appears, proved much more efficient. The simplest of these systems can indeed be regarded as only a formalization and extension of the inarticulate model-mimic arrangement. This is the apprenticeship system, in which teaching is still largely by showing and only partially in the form of words and formal instructions. Human apprentice teaching, however, differs from animal model-mimic transmission in such factors as the conscious utilization of repetition as a means of indoctrination, and in its development of a long-term course of instruction leading towards a definite recognized goal. It is still utilized even in the most highly evolved societies at many levels of sophistication, from the training of a plumber or carpenter through that of a doctor or lawyer to the most rarified spiritual level of a guru and his pupil.

The next step in the evolution of sociogenetic transmission mechanisms is, perhaps, the formalization of rote learning. In general, in modern societies rote learning is used to inculcate information which has somewhere or other been recorded in written form. This is the case, for instance, in the rote learning of the Koran or the Confucian classics which are features of classical Mohammedan and Chinese education. The existence of a written text is, however, not a necessary adjunct of rote learning. One might hazard the guess, for instance, that the bards of Homer's day, like the court singers of classical Ireland, Iceland and West Africa, to name a few cultures at random, were taught largely by rote with an almost total absence of any writing.

PROGRESS BEYOND PERSON-TO-PERSON
TRANSMISSION OF LEARNING

It was, of course, the invention of writing which removed from the sociogenetic mechanism the necessity for person-to-person contact between the transmitter and the recipient. It is scarcely necessary, and certainly impractical in a lecture such as this, to elaborate the steps in the development of techniques of social transmission through recorded conceptual language. I will only make the remark—not entirely a facetious one—that impressive though these advances have been in some respects, we seem in some ways only too reluctant to take advantage of them. Why, the University teacher often thinks, must he continue to spend so much of his time using for the nth year in succession a technique of instruction which was sensible enough for his forefathers who had no alternative method available except manuscript treatises inscribed on vellum? Whereas, he nowadays could so easily record, not only his voice, but if you wish an accurate representation of his expression and gestures, on a piece of magnetic tape, while he himself wrote learned works in a villa in the south of France or conducted experiments in his laboratory.

It is characteristic of all the more highly evolved means of sociogenetic transmission that they do not depend essentially on person-to-person contact. The biological genetic system has of course never escaped this limitation. Every individual animal must have an individual mother and the majority of them an individual father also. The major weakness of the biological system from an evolutionary point of view is that, although it ensures some mixing of hereditary qualities from different individuals, it limits the number of individuals which can participate in this mixing to only two; and in practice it usually operates to reduce the difference between these two since it is difficult to persuade mother and father of widely different kinds to hybridize. This tendency against hybridization is very greatly reduced in the human sociogenetic system. One of the major features of human evolutionary processes, in fact, is the incorporation into one culture of elements which have arisen in another. The process is comparable to what is known in the botanical world as introgressive hybridization. In the biological realm the results from a wide hybridization usually differ from the original type not so much by presence or absence of individual genes, but rather by the persistence or loss from the strain of whole chromosomes. In the sociogenetic system of man also, when cultures come in contact it seems that what they take from each other is not a number of separate discrete items, but rather large portmanteau chunks of information. It is not, I think, quite clear whether these chunks are comparable to chromosomes; that is to say, are composed of a more or less randomly-associated set of items which happen to occur together, as do the genes in a linkage group on the chromosome; or whether the chunks are always more highly organized in a manner comparable to the epigenetically-interactive groups of gene activities which characterize the different tissues of an animal body. When one culture adopts from another, say, a religion such as Christianity, it is clear that we are dealing with a group of items which is organized and which is comparable rather to an epigenetic grouping than to a purely genetic linkage group. On the other hand, when the Indian culture adopts from the British such diverse items as a taste for cricket, a particular method of parliamentary election, a certain organization of the army, and a special administrative jargon such as 'I beg you to do the needful', it seems rather more likely that we are dealing with the introgression of an arbitrarily associated group of items, comparable to the genes in a chromosome.

The varied items of information which are transmitted by these different mechanisms must have originated in some way when they first began their social career. It has been suggested above that in the sociogenetic system incorporation of items from other cultures (processes comparable to introgressive hybridization), and the transmission of complexes of units whose unity is essentially epigenetic, are both much more frequent and important processes than the comparable happenings in the subhuman biological system. But they cannot be the whole story of the origin of variation in the human cultural heritage. There must be some process by which new items of socially transmissible information are added to the human store. In biological evolution much of the genetic variation in a population is created anew in each generation by recombination of already existing genes, but we fully realize that this shuffling is essentially a second-order process, based on a primary process of mutation by which new alleles are created. What is the comparable process to mutation in the sociogenetic system?

The answer must be two-fold. In the first place, one may point out that in sociogenetics a process comparable to the inheritance of acquired characters undoubtedly occurs. Cultural transmission, as we have seen, does not depend on specialized transmitting entities, comparable to gametes, but the situation is as though any differentiated tissue of the animal's body could transmit its qualities direct to the next generation. Thus, much of the new variation in sociogenetically transmitted items can arise as acquired characters, from the interaction between the human beings in the population and their surroundings. The human evolutionary system can thus utilize a vast source of variation which is more or less closed to the sub-human world. But it is by no means certain that all new sociogenetically transmittable items arise in this way. Man appears to develop ideas whose nature is not a necessary consequence of the environmental circumstances, and in so far as this is the case, these ideas can scarcely arise solely as acquired characters. It is difficult, for instance, to deny that there is some arbitrary element in the distinction between the great religions, such as Christianity and Buddhism. There would seem to be a place in the genesis of new human ideas for some process which shares with gene mutation a characteristic of randomness and unpredictability.

We have recently come to realize that a very important property of the biological genetic system is the existence within natural populations of a tendency which has been called genetic homeostasis. A natural population of animals contains a pool of genes, the particular genetic endowment of any one individual being a sample drawn out of this pool. If some disturbing agency, such as natural selection, is applied to the population, the frequency of the various genes in the pool will be altered. The tendency of the population to genetic homeostasis is exhibited by the fact that when the selection pressure is released the frequencies of the various genes frequently return towards, or even reach, their original values. The genetic make-up of the population, in fact, exhibits some resistance to agents which would tend to change it. A similar balance between a certain degree of flexibility combined with some resistance to modifying agents is, as we have pointed out above, also shown by the epigenetic systems which lead from the egg to the adult condition.

QUESTIONS TO THE SOCIOLOGIST

In the sociogenetic system there are parallels for both these types of qualified stability. In human culture, however, it is not so easy to distinguish between them. As we have seen, the sociogenetic system includes phenomena comparable to the inheritance of acquired characteristics and the transmission of epigenetically organized complexes. Thus there are several mechanisms by which a human culture can manifest resistance to change, but it is not easy to classify them in groups comparable to those which are applicable to the biological evolutionary system. One agent opposing change may be the mechanism of sociogenetic transmission itself: for instance, many cultural traits which are conveyed by the primitive inarticulate transmission mechanism, which was discussed first above, seem to be very resistant to change, probably because of the transmission filter through which they have to pass. Examples are the somewhat nebulous, but often easily recognizable, qualities which are often spoken of as national characteristics, such as Jewishness or Indianness. Here, perhaps the sociogenetic mechanism is producing something closely comparable to biological genetic homeostasis in its strict sense. At the other end of the range there are tendencies to stability in human culture which arise from what is clearly an epigenetic unity; for instance, a closely organized body of dogmas such as that of the Roman Catholic Church is not at all easy to alter and tends to be transmitted more or less unchanged from generation to generation. In between these two extremes there are very many intergrades. It does not seem likely that much purpose would be served by trying to classify them on the basis of their logical similarity to comparable genetic and epigenetic phenomena in biology.

There is in this general field one consideration about human society which does, I think, raise some interesting questions. Most animals retain only for a very short fraction of their lifespan any capacity to be modified by their environment. The period in which they are epigenetically flexible is comparatively short. Man can—indeed in the present century probably must   go on learning throughout practically his whole life. One might then institute a comparison between a whole generation within a given human culture and a single animal individual, regarding the changes which the human generation undergoes during its lifetime as comparable to the changes which the egg will undergo as it develops into an adult. One would find then, I think, that what we may call the 'socio-epigenetic' system of a generation varies considerably from culture to culture. In some there is much more resistance to change during a given lifetime than in others. In American culture, for instance, it is a matter of pride that an individual should adopt new habits and modes of life as they come along. In Britain, classical China, and many other countries much more value is attached to clinging to the old ways. This is very comparable to the fact that in animal species one can find some whose development is extremely resistant to modification by the environment—is, in fact, what has been called strongly canalized—while in others the epigenetic processes are much more flexible. For instance, mice inhabit a very large variety of habitats, but look very much the same in all of them, whereas some invertebrate species are so easily modifiable that almost every pond in which they live has its own recognizable population. In the biological realm the strength and character of the epigenetic canalization can certainly be controlled by selection, and is a factor which plays an important role in evolution. In the sociogenetic system the readiness to accept change within a generation must also be both influenced by, and itself influence, the evolutionary processes.

Closely allied with these matters is the problem which Darwin made central in his work—the problem of speciation. It is an empirical fact that living organisms do not vary continuously over the whole range which they exhibit, but that they fall into more or less well defined groups, which are commonly called species. The precise definition of what constitutes a species is a matter of great difficulty, about which biologists are hardly yet in agreement, but that some significant discontinuity occurs can scarcely be questioned. Paradoxically enough, however, the origin of species is just the facet of evolution on which Darwin's theories throw the feeblest illumination. We still have very little understanding of why discontinuity occurs so frequently. We have to suppose that in some way certain constellations of hereditary potentialities fit together into a stable pattern, while other combinations are inharmonious; but that is only a very abstract and general statement. But, although biologists do not understand it, here is an area of enquiry which, one feels, sociologists will also have to face. To what extent is discontinuity a characteristic of the variation between human cultures? To the outsider it would seem that we find phenomena extremely similar to those with which the biologist is familiar; some instances of sharp distinctions between even closely neighbouring cultures, as in such a culturally diversified area as New Guinea; some examples of more or less continuous geographical variation, comparable to the formation of local races, in widespread cultures such as, to take an extreme example, the British with its offshoots in Canada, New Zealand, Australia, etc.; and a tendency, also familiar in biology, for the initial slight geographical variants to evolve into fully distinct 'species'. The dynamics of this process—for instance its dependence on, or independence of, the formation of barriers to cross-mating—would appear to present sociologists with problems very similar to their formal structure at least to those with which biologists are wrestling.

Finally, I think the biologist would wish to ask the sociologist whether there is in his system of ideas anything which plays the evolutionary role of natural selection. Something, after all, must decide which new items of culture, either adopted from other societies by processes akin to hybridization or arising 'out of the blue' by some analogue of mutation, will succeed in persisting for many generations. Why, for example, were Christ and Mohammed accepted as Messiahs out of all the candidates for that role? Or why were Lamarck's and, for a long time, Mendel's ideas rejected while Darwin's won immediate acclaim? The processes are, perhaps, so complicated that we cannot hope to find any general portmanteau term like 'Natural Selection' to apply to them. But one wonders whether this is not too pessimistic a view. The natural selective value of a new biological variant depends on the number of its offspring. It is possible that the ability of a new cultural item to persist could be deduced from the magnitude of its cultural progeny—its ability to 'cross-breed' with already existing facets of the culture and to beget issue from them. And of course we should not forget that many biologists now attribute a considerable influence to the fluctuations of random sampling as a phenomenon which mitigates the rigours of strict natural selection. How great a role does pure chance play in the preservation or disappearance of new cultural items?

In making these remarks about human evolution it has not been my intention to suggest that our knowledge about the biological evolutionary system will enable us to answer the problems with which sociologists are confronted. All I have tried to show is that a number of quite interesting lines of thought emerge if one takes in turn each of the factors which we consider important in biological evolution, and asks oneself what corresponds to them in the human cultural system. Darwinism in its heyday seemed to some of its enthusiastic supporters to have all the answers on the biological level, and they were so convinced of this that they felt that it must also provide the answers in sociology. Nowadays, even within biology I think we are more modest. We may feel that Darwinism, as it has been modified and developed in the last hundred years, provides us with the main principles which we require for an understanding of biological evolution, but perhaps the most striking feature of the last two or three decades has been the realization of how much still remains to be discovered. There are probably rather few biologists —though there certainly seem to be some—who would be confident in asserting that their science gives the essential key to the understanding of human affairs. But the modes of analysis used by the biologist may be at least suggestive to sociologists, and I hope that this attempt to look at their field from the angle of recent evolutionary theory will provoke thought, even if the result of that cogitation is to reject all the suggestions I have made. [1961]

29

THE EVOLUTION OF ALTRUISM AND LANGUAGE

Most of the general theory of evolution is concerned with characters whose advantages or disadvantages for natural selection affect only the individuals which bear them, or perhaps their immediate sexual partners. There are, of course, other kinds of character, which might be called 'social', since their natural selective effects either depend on or influence the relationships between the individual exhibiting the character and a more or less large number of other members of the same species. The evolution of such characters presents some special problems.

In this essay, three classes of social characters will be considered; altruistic behaviour, conventional behaviour, and language.

Altruistic behaviour is that which is useful to the social group to which an animal belongs. Haldane[1] was the first to draw attention to the difficult problem of explaining the evolution of types of altruistic behaviour which are harmful to the individual who behaves in this manner. Clearly, evolution in this case cannot depend on direct selection of the altruistic individuals; it must depend on processes operating at the level of the entity whose selective value has been improved, namely, the group. Haldane argued that this could only be effective in a species in which the population is organized into rather clearly separated groups, between which there is little movement of individuals; and, even in such circumstances altruistic behaviour would not spread unless it was either extremely effective in improving the natural selective value of the group or unless it was exhibited by a large proportion of the individuals. However, some years later, Haldane pointed out[2] that many of the difficulties disappear if one remembers that the members of a group who are assisted by an altruistic individual will be genetically related to him, and will therefore carry many of the same genes. An individual who sacrifices himself for his group will not necessarily be reducing the chance that his genes are represented in the next generation; he may well be increasing it. Haldane gives the following illustration (quoted in[3]):

> You can carry a rare gene which affects your behaviour so that you jump into a flooded river and save a child, but you have one chance in ten of being drowned. . . . If the child is your own child or your brother or sister, there is an even chance (on Mendelian principles) that this child will also have this gene, so five such genes will be saved in children for one lost in each adult. If you save your grandchild or nephew the advantage is only $2\frac{1}{2}$ to one. If you only save a first cousin, the effect is very slight. If you try to save your first cousin once removed, the population is more likely to lose this valuable gene than to gain it.

The name 'kin selection' has been proposed by Maynard Smith for this type of process. It has been quite thoroughly discussed recently[3,4,5,6,7] and there seems no reason to doubt that it provides a satisfactory explanation for the evolution of altruistic behaviour in most situations in which that is found.

Another type of social behaviour in animals, which has always been regarded as presenting particular difficulties for evolutionary theory, is that in which animals appear to co-operate to reduce the severity of some types of natural selection. For instance, in many species, individuals of one sex, usually males, compete directly with each other for access to breeding females or for territorial or other resources, which are of obvious importance in determining the number of offspring they will leave. Often this 'competition' takes the form of actual fighting, but, in a large majority of species, these fights are not pushed to the point at which the winner inflicts serious damage on the loser. Why not? Observation of animals in nature shows that in most cases, the loser performs some action or gesture

which is recognized by the winner as 'throwing in the towel', or con-
ceding the contest, and the winner holds back and lets the loser
escape. The fight is controlled by some sort of socially recognized
convention.

It seems at first sight that it would be in the interests of any
individual always to fight as hard as he could, to make sure of winning if at all possible. People who have looked at the situation in that
way, have pointed out that, although such a course might be good
for the individual, and would therefore be favoured by natural
selection within any population, it might be bad for the group as a
whole, by leading to too many deaths, among the young and fairly
vigorous males, leaving only the few very good fighters. They have
argued that if a population is divided into a number of smaller
groups, with little cross-breeding between them either because they
occupy different territories or for some similar reason, but if these
occasionally do come into competition with one another, then a
group which has adopted a convention to limit the damage inflicted
on it by internecine fights, may prove to be stronger than a group
which has not done so, so that it wins the 'inter-group competition'.
It has been suggested that the social conventions have been evolved
by this mechanism of 'group selection' rather than by the selection
of individuals with which most evolutionary theory is concerned.

There seems no reason to doubt that such inter-group selection
could occur; it certainly does often enough in the human species, as
when the Europeans arrived in Australia or North America. But the
conditions under which it can occur are rather special: in fact it is
easy to find one is advancing a circular argument, which has to
postulate strong forces of social integration to hold together the
groups whose competition is being appealed to as an explanation of
the evolution of socially integrating conventions. It would certainly
be comforting if one could find some other plausible arguments,
preferably arguments relying on the selection pressures on individuals
which are the mainstay of most other evolutionary processes.

Actually, it is not at all difficult to think of other processes than
group selection which might have brought about the evolution of
these social conventions. The argument that it would always be to the
individual's advantage to carry a fight to its bitter end really rests on
a number of quite questionable assumptions—after all, it conflicts
with several pieces of traditional wisdom: 'He who fights and runs
away, Lives to fight another day'; and 'Discretion is the better part
of valour'. The argument in its simplest form assumes that each pair
of males engages in only one fight, and that the winner can take all.

It becomes much more difficult to argue through the consequences if what is being competed for is such that even the winner could not take all of it, or if he must expect that after winning his first fight he will be challenged again by another competitor. Then the extent of the spread of fighting abilities within the population, and where any particular individual stands in the rank order will have important effects on the amount he stands to win or lose from various possible types of behaviour.

Maynard Smith has begun the task of working out a mathematical model for discussing these problems, using the methods of Games Theory[6,8]. He considers three types of behaviour; 'dangerous' fighting, carried far enough to inflict serious damage on the loser; 'conventional' fighting, which is not dangerous; and 'retreat', breaking off the combat. He considers what will happen in a combat in which the two contestants employ various sequences of these behaviours, and of course may change from one mode to another in response to what the other animal is doing. In order to evaluate the final result of a fight, he has to assign 'pay-offs', which put a numerical value on winning or losing (and he also allows for the waste of energy in taking part in a long fight). Of course these pay-offs are not empirically known; but in order to explore the theory of the situation, it is reasonable to give them various arbitrary values. Maynard Smith used a computer to work out the consequences of various sequences of behaviour in large numbers of 'fights', which were varied by feeding in pseudo-random numbers. He looked for a strategy such that, in a population in which the great majority of members acted in that way, no other strategy would give better results. This he calls an 'Evolutionary Stable Strategy', or ESS.

Which type of behaviour does turn out to be an ESS depends critically on the values chosen for the various pay-offs. For instance, if an animal suffers as much damage if he loses a fight, even retreating before getting hurt, as he would do if he fought to the bitter end and lost, then it turns out that the stable strategy is one in which all individuals fight as hard as they can. But the important result at which Maynard Smith arrives is that there are some quite plausible pay-off values which result in strategies involving conventional ways of 'throwing in the towel' becoming evolutionarily stable. This is a theoretical demonstration that if a species had once arrived at a state in which most individuals exhibit the 'limited warfare' type of behaviour, then it requires only considerations of individual selection to keep it in that condition, and it is not necessary to appeal to group selection.

At the present time, this conclusion must be regarded as a very interesting first approach, rather than a fully satisfactory theory. To reach it Maynard Smith has made a number of simplifications, the importance of which is not yet fully understood. For instance, he has supposed that all individuals are equally effective in the actual fighting. It is not quite obvious what would happen in a population in which there were much better fighters than others, so that they almost always win if they went all out with no reference to conventions. Indeed Maynard Smith does not explicitly consider individuals who engage in sequences of fights, although he points out that some of the results of this can be taken account of in the pay-off assigned to losing. For instance, the situation mentioned above, where even a conventional loss is as damaging as a real loss, corresponds to a situation where an animal fights only once, perhaps for a mate, and never gets a second chance. But the possibilities need a good deal more exploration than Maynard Smith has yet been able to give them.

The main defect of the theory, as it exists at present, is that it is in terms of strategies which are evolutionary stable, but tells us very little about how a population could come to adopt such a strategy in the first place. This involves two rather different questions. The first is to enquire what will happen in a population in which only a proportion of the individuals adopt a conventionalized signal for terminating a combat. One must remember that, to be effective, such signals have not only to be given, but also received and acted on. I think one's first instinctive thought would be that the convention would not be of advantage to its practitioners until they amounted to a sizeable fraction of the population. If that were so, it is difficult to see how they ever reach that situation. But I do not think the problem has been properly studied as yet.

The second question is, how did the conventional signal come to be selected, out of all the possible items of behaviour. In the conventions observed in animal fighting, the signal is usually some behaviour which is obviously unsuitable for carrying on the fight, for instance, rolling on the back and offering a soft, unguarded underbelly; but there might be several actions all equally a 'give away'. And in the ritualized combats involved in competitive displays how did it come about that evolution has fixed on the particular routines of tail spreading, strutting, pushing out the breast, or ruffling the neck feathers, which are characteristic of different species? Of course it is usual to find that the parts of the animal used in the display have unusual or striking patterns, like the peacock's tail; but is it not just a circular argument to say that that is why they are used, and the fact that they are so used is why they have evolved their special features? But perhaps evolution, which is a very gradual process, progressing by many small steps, is less averse to circular arguments than is our abstract logic.

The selection of an item of behaviour to act as a conventional sign, part of a system of communication with another individual, can perhaps be regarded as a first evolutionary step towards one of the most complex and certainly one of the most important of all social characters: the ability to use language. It was for a long time thought that the human species stood completely alone in the living world in its ability to communicate by the use of symbols in the form of words. It is gradually becoming apparent that, although man still seems in a class by himself in the efficiency and flexibility of his use of language, some other animals, and particularly the higher apes most closely related to man, show at least traces of somewhat similar abilities. (Enthusiastic claims that Dolphins have a language as rich as the human have not stood up to more careful examination). Any believer in the theory of evolution would, of course, expect there to be some signs of the rudiments from which a fully evolved language might have been built up. What is surprising is the magnitude of the gap between the human ability and the best that any other animal can show in this direction.

The subject is under very active investigation and discussion by specialists in ethology and linguistics. A recent summary of many of the views which are most important at the present time will be found in Hinde[9], and this also includes a very extensive bibliography; but anyone who wishes to examine the full range of present ideas should supplement this by the study of Suzanne Langer's book[10], especially volume II, which only appeared in 1972, probably too late to be referred to in the volume edited by Hinde. Langer attaches a much greater importance, in the genesis of language, to the attachment of simple words, at that stage little more than humanly produced noises, to emotions which came to be socially shared, rather than directly to objects; although she, of course, agrees that at a fairly early stage most of the symbols would have their reference transferred from the sheer emotion to some object associated with it. However, it is clear that this topic is a very complex one, and it would be foolish to try to discuss the general principles involved in a short essay. My purpose here is only to indicate ways in which it seems that the point of view about evolutionary processes which has been developed in these essays may have a bearing on one or two aspects of the language problem which have seemed to some to raise difficulties for conventional evolutionary theory.

The first of these difficulties is that which was referred to above, namely the enormous gap between the abilities of man and any other animal in the use of language. If language had evolved by any Darwinian process it must have reached its present perfection through a series of gradual improvements, and if so, why do we not find any trace of them? One possible answer might be that we are dealing here with a case of evolution depending on the appearance of a large discontinuous change in hereditary property, on a 'mutation' in the sense in which that word was interpreted by the first geneticists at the beginning of this century, such as de Vries or Bateson. And another reason has been put forward for invoking the occurrence of some genetic mutation of profound effect in relation to language. One of the most influential of recent students of linguistics, Noam Chomsky [11], has argued that the ability of children to learn language is so remarkable, and that the structures of all known languages have such a remarkable uniformity in underlying grammatical structure, that one has to postulate that there is innate in the human species, that is genetically encoded in its genome, a capacity to operate languages having such structures. For Chomsky, the acquisition of a language is more a matter of the activation of this genetically determined ability than a process in which *everything* has to be learnt.

But, as we have seen, genetical students of evolution have long ago given up relying on the occurrence of mutations which are 'major' in the sense that they produce drastic changes in phenotype and radically novel characteristics. Certainly the whole of statistical neo-Darwinism is in terms of mutations which have quite minor phenotypic effects, which contribute to a more or less continuous range of variation. The epigenetic or post-neo-Darwinist theories which I have been advocating invoke mainly genes with even slighter phenotypic effects, often in fact producing no phenotypic change at all except in cooperation with certain particular environmental factors. However, although the neo-Darwinist view does, I think, make it very difficult to see how the kind of innate language-using ability invoked by Chomsky could come into being, I think the post-neo-Darwinist view might help to solve this problem (see also [12]).

In the first place, let us return to the question of the magnitude of the gap between the human and other species in language use. It is an example, quite strong though I think not extreme, of a type of problem which arises in many fields of evolution. We have, for instance, very few signs of the evolutionary processes leading from the reptiles to the birds, and, of course, no signs of the evolutionary gaps between many of the invertebrate phyla. In all cases like this, I think one can invoke the principle of archetypes; evolution responding to some moderate changes in natural selective pressure may proceed through small gradual steps for a considerable time, and then may suddenly produce a form which has a stability, or a capacity of response, of a different order of magnitude to anything which has been demanded of it in the past. It may either meet an old challenge so very efficiently that it rapidly and completely takes over from all the old and much less efficient phases that have led up to it; or it may be capable of being used in some radically new way which allows the ancestral forms to survive only in much simpler contexts where their ancestral relation to the novelty may be almost unrecognizable. Language which developed to the human state may well be such an archetypal novelty. It would be such an incomparably more effective means of communication than any system which had a few but not all its major properties, that any such evolutionary intermediates would have been very rapidly supplanted; in such circumstances one could hardly expect to find anything of them surviving to the present day. One would expect either languages which had human capacity, or the most primitive systems of semi-symbolic, basically agrammatical grunts, squeaks etc. which one could hardly recognize as being languages at all; and that is just what we do find.

The effectiveness of what I have called an archetype does not depend on the occurrence of any 'macro-mutation'. Its properties arise not from the magnitude of the hereditary change, but from the character of the organization which develops within the phenotype. The epigenetic theory of evolution also suggests how we can surmount the difficulty of accounting for Chomsky's innate language ability without invoking a macro-mutation. This could be simply an example of genetic assimilation. Any theory of the evolution of language must postulate that language use is selectively valuable, and this is surely not an unreasonable assumption, whether primitive languages were used for communication between members of hunting parties or, as Suzanne Langer suggests, in rituals which increase the social cohesion within a group of individuals. If there were selection for the ability to use language, then there would be selection for the capacity to acquire the use of language, in an interaction with a language-using environment; and the result of selection for epigenetic responses can be, as we have seen, a gradual accumulation of so many genes with effects tending in this direction that the character gradually becomes genetically assimilated.

In the characters which have been studied in the laboratory, which of course do not include language, the genetic assimilation has gone so far that in later populations the character can be developed in the absence of what initially was a necessary environmental stimulus. Genetic assimilation of language has presumably not gone so far. It has been reported that King James VI of Scotland and I of England did the experiment of confining two children of a very early age on an isolated island in the Firth of Forth to see what language they would speak, and duly found, of course, that they grew up to speak ancient Hebrew, the language of the Garden of Eden. More recent, and I am afraid more reliable if less picturesque, authors agree that human individuals brought up in the absence of contact with language-speaking people (and a few such odd cases are known), do not develop any language spontaneously. What is asserted by Chomsky and his followers is that the human individual develops many of the mental capacities (i.e. presumably neurological mechanisms) necessary for the use of language, at a stage earlier than he actually begins to put them to that use. He does not bring to his first experiences of language a mental apparatus which is either completely without any built-in mechanism, or one containing a great range of possible mechanisms—neither a heap of cogs, pistons, springs and so on, nor a flexible workshop containing lathes, drills, milling tools, internal combustion engines, electric motors, hydraulic presses and so on; but rather his mind contains certain rather definite capacities for handling symbolic communications systems of a particular kind in particular ways. It is this particular mental apparatus which I suggest might have been built up by a process of genetic assimilation. There could be a further step in such a process, and if the assimilation went far enough, this machinery could start operating in an environment from which other language-users were quite removed. Evolution is quite capable of performing such a feat, as it does, for instance, in a skin on the soles of our feet, or the callosities of the ostrich, which put in an appearance in the right places on the embryo before it has escaped from the womb or the egg shell. But in the case of language, there is certainly little reason to see why it would have been advantageous to press the matter so far. If a child which had never met a language-user developed the ability to talk, who after all would it have to talk to? [1974]

Notes and References

PAPER 1 (1–11)

1 The Second Symposium on 'Towards a Theoretical Biology', held at the Villa Serbelloni, Lake Como, Italy. Proceedings published by Edinburgh University Press 1969.
2 This was a slip; actually the shell was that of a lamellibranch, *Avicula dorsetensis*.

PAPER 2 (11–15)

1 GOLDSCHMIDT,R. (1927) *Physiologische Theorie der Vererbung*. Berlin: Springer.
2 WADDINGTON,C.H. (1940) *Organisers and Genes*. Cambridge: Cambridge University Press.
3 GOLDSCHMIDT,R. (1940) *The Material Basis of Evolution* (Mrs Hepsa Ely Silliman Memorial Lectures). New Haven, Conn.: Yale University Press, and London: Oxford University Press.
4 WILLIS,J.C. (1940) *The Course of Evolution by Differentiation or Divergent Mutation rather than by Selection*. Cambridge: Cambridge University Press.

PAPER 3 (16–22)

1 ROBSON,G.C. & RICHARDS,O.W. (1936) *The Variations of Animals in Nature*. London.
2 DUERDEN,J.E. (1920) The inheritance of the callosities in the ostrich. *Amer. Nat.*,54,289.
3 DARWIN,C. (1901) *The Descent of Man and Selection in Relation to Sex*. London.
4 DETLEFSEN,J.A. (1925) The inheritance of acquired characters. *Physiol. Rev.*,5,244.
5 HUXLEY,J.S. (1942) *Evolution: the Modern Synthesis*,74. London.
6 PLUNKETT,C.C. (1932) Temperature as a tool in research in phenogenetics. *Proc. 6th int. Congr. Gen.*,2,158.
7 FORD,E.B. (1940) Genetic research in the Lepidoptera. *Ann. Eugen.*,10,227.
8 FISHER,R.A. (1928) The possible modification of the response of the wild type to recurrent mutations. *Amer. Nat.*,62,115.
9 STERN,C. (1929) Uber die additive Wirkung multipler Allele. *Biol ₵bl.*,49,231.

10 MULLER,H.J. (1932) Further studies on the nature and causes of gene mutations. *Proc. 6th int. Congr. Gen.*, **1**, 213.
11 WADDINGTON,C.H. (1940) *Growth* Suppl., 37.
12 MATHER,K. & DE WINTON,D. (1941) Adaptation and counter-adaptation of the breeding system in *Primula. Ann. Bot.*, **5**, 297.
13 KUHN,A. (1936) Versuche über die Wirkungsweise der Erbanlagen. *Naturwiss*, **24**, 1.
14 WADDINGTON,C.H. (1940) *Organisers and Genes* (Cambridge); Genes as evocators in development.

PAPER 4 (22–4)
 1 DOBZHANSKY,TH. (1937) *Genetics and the Origin of Species.* Columbia University Press.
 2 WADDINGTON,C.H. (1942) *Nature*, **150**, 563.

PAPER 5 (24–35)
 1 MEDAWAR,P.B. (1951) *New Biol.*, **11**, 10.
 2 HARRISON,R.G. (1929) *Arch. EntwMech. Org.*, **120**, 1.
 3 HADDOW,A. (1947) *Brit. med. J.*, **4**, 417.
 4 WEISS,P. (1947) *Yale J. Biol. Med.*, **19**, 235.
 5 EBERT,J.D. (1950) *J. exp. Zool.*, **115**, 351.
 6 DRASTICH,L. (1925) *Z. vergl. Physiol.*, **2**, 632.
 7 SEMON,R. (1913) *Arch. mikr. Anat.*, **82**, 164.
 8 DUERDEN,J.E. (1920) *Amer. Nat.*, **54**, 289.
 9 LECHE,W. (1902) *Biol. Zbl.*, **22.**
10 KUKENTHAL,W. (1892–1912) in *Zoologische Forschungsreise in Australien*, (ed. Semon, R.) vol. v. Jena: G. Fischer.
11 WADDINGTON,C.H. (1940) *Organisers and Genes.* London: Cambridge University Press.
12 WADDINGTON,C.H. (1942) *Nature*, **150**, 563.
13 WRIGHT,S. (1935) *Genetics*, **20**, 84.
14 WADDINGTON,C.H. (1953) *Symp. Soc. exp. Biol.*, **7**, 186.
15 WADDINGTON,C.H. (1953) *Evolution*, **7**, 118.
16 Discussion on the variation of *Lymnaea. Proc. malacol. Soc. Lond.*, **23**, 303, 1939; see also paper 9 in this collection.

PAPER 6 (35–6)
 1 SCHROEDINGER,E. (1954) *Nature and the Greeks.* Cambridge: Cambridge University Press.
 2 WADDINGTON,C.H. (1942) *Nature*, **150**, 563; (1953) *Evolution*, **7**, 118, 386.
 3 WADDINGTON,C.H., and others (1941) *Nature*, **148**, 270; (1942) *Science and Ethics.* Allen and Unwin.

PAPER 7 (36–59)
 1 ROE,A. & SIMPSON,G.G., eds. (1958) *Behaviour and Evolution.* New Haven: Yale University Press.
 2 DURRANT,A. (1958) Environmentally induced inherited changes in flax. *Proc. 10th int. Congr. Gen.*, **1**, 71.

3 HIGHKIN,H.R. (1958) Transmission of phenotypic variability in a
   pure line. *Proc. 10th int. Congr. Gen.*,2,120.
4 HESLOP-HARRISON,J.W. (1920) Genetical studies in moths.
   J. Genet., IX,195.
5 KETTLEWELL,H.B.D. (1955) Selection experiments in industrial
   melanism in the Lepidoptera. *Heredity*,IX,323.
6 MAYR,E. (1958) Behaviour and systematics. *Behaviour and Evolution*
   (eds. Roe,A. & Simpson,G.G.),341. New Haven: Yale University
   Press.
7 WADDINGTON,C.H. (1959) *Nature*,CLXXXIII,1654.
8 GLOOR,H. & CHEN,P.S. (1950) Ueber ein Analorgan bei
   Drosophila Larven. *Rev. suisse zool.*,LVII,570.
9 TURESSON,G. (1930) The selective effect of the climate upon plant
   species. *Hereditas*,XIV,99.
10 WADDINGTON,C.H. (1940) *Organisers and Genes*. London:
   Cambridge University Press.
11 WADDINGTON,C.H. (1942) The canalization of development and
   the inheritance of acquired characters. *Nature*,CL,563.
12 WADDINGTON,C.H. (1953) The genetic assimilation of an acquired
   character. *Evolution*,VII,118.
13 BATEMAN,G. (1956) Studies on genetic assimilation. PHD thesis,
   University of Edinburgh. *J. Genet.*, 56 (1959) 341–52 and 443–74.
14 WADDINGTON,C.H. (1956) Genetic assimilation of the bithorax
   phenotype. *Evolution*,X,1.
15 WADDINGTON,C.H. (1957) The genetic basis of the assimilated
   bithorax stock. *J. Genet.*,LV,241.
16 GLOOR,H. (1947) Phaenokopic Versuche mit Aetter an Drosophila.
   Rev. suisse zool.,LIV,637.
17 WADDINGTON,C.H. (1957) *The Strategy of the Genes*. London: Allen
   and Unwin; New York: Macmillan Co.
18 KNIGHT,G.R., ROBERTSON,A. & WADDINGTON,C.H. (1956)
   Selection for sexual isolation within a species. *Evolution*,X,14, and
   paper 15 in this collection.
19 KOREF,S.S. & WADDINGTON,C.H. (1958) The origin of sexual
   isolation between different lines within a species. *Evolution*,XII,485,
   and paper 16 in this collection.
20 SPEITH,H.T. (1958) Behaviour and isolating mechanisms.
   Behaviour and Evolution (eds. Roe,A. & Simpson,G.G.),363. New
   Haven: Yale University Press.
21 WADDINGTON,C.H., WOOLF,B. AND PERRY,M.M. (1954)
   Environment selection by *Drosophila* mutants. *Evolution*, VIII,89,
   and paper 14 in this collection.
22 KETTLEWELL,H.B.D. (1956) Investigations on the evolution of
   melanism in Lepidoptera. *Proc. Roy. Soc. Lond. B.*,CXLV,297.

PAPER 8 (59–92)

1 WADDINGTON,C.H. (1953) Genetic assimilation of an acquired character. *Evolution*, **7**,118–26.

2 GOODRICH,D.S. (1934) *Living Organisms*. London and New York: Oxford University Press.

3 BEGG,M. (1952) Selection of the genetic basis for an acquired character. *Nature*, **169**,625.

4 WADDINGTON,C.H. (1952) Reply to Begg (1952). *Nature*, **169**,625.

5 WADDINGTON,C.H. (1942) Canalization of development and the inheritance of acquired characters. *Nature*, **150**,563–4.

6 WADDINGTON,C.H. (1951) The evolution of developmental systems. *Rept. Australian and New Zealand Assoc. Advance Sci.*, **28**,155–9.

7 WADDINGTON,C.H. (1953) Epigenetics and evolution. *Symposia Soc. Exptl. Biol.*, no. 7,186–99.

8 WADDINGTON,C.H. (1954) The integration of gene-controlled processes and its bearing on evolution. *Proc. 9th int. Congr. Genet.* (Caryologia Suppl.) **6**,232–45.

9 WADDINGTON,C.H. (1957) *The Strategy of the Genes*,262. London: Allen and Unwin.

10 WADDINGTON,C.H. (1958) Inheritance of acquired characters. *Proc. Linnean Soc. London*, **169**,54–61.

11 WADDINGTON,C.H. (1958) Theories of evolution. *A Century of Darwin* (ed. Barnett,S.A.),1–18. London: Heinemann.

12 WADDINGTON,C.H. (1959) Evolutionary systems: animal and human. *Nature*, **183**,1634–8.

13 WADDINGTON,C.H. (1959) Evolutionary adaptation. *Evolution after Darwin, I. The Evolution of Life.* (ed. Tax,S.),381–402. Chicago: Chicago University Press; see also *Perspectives in Biol. Med.*, **2**, and *Vestnik Ceskoslov. Zool. Spol.*, **23**,289–306.

14 WADDINGTON,C.H. (1952) Selection of the genetic basis for an acquired character. *Nature*, **169**,278, and paper 4 in this collection.

15 BATEMAN,K.G. (1956) Studies on genetic assimilation. PhD thesis, University of Edinburgh. *J. Genet.*, **56** (1959) 341–52 and 443–74.

16 BATEMAN,K.G. (1956) Studies on genetic assimilation. *Proc. Roy. Phys. Soc. Edinburgh*, **25**,1–6.

17 BATEMAN,K.G. (1959) Genetic assimilation of the dumpy phenocopy. *J. Genet.*, **56**,341–52.

18 BATEMAN,K.G. (1959) Genetic assimilation of four venation phenocopies. *J. Genet.*, **56**,443–74.

19 MILKMAN,R. (1955) Unpublished PhD thesis and personal communication.

20 WADDINGTON,C.H. (1956) Genetic assimilation of the bithorax phenotype. *Evolution*, **10**,1–13.

21 WADDINGTON,C.H. (1957) The genetic basis of the assimilated bithorax stock. *J. Genet.*, 55, 241–5.

22 WADDINGTON,C.H. (1959) Canalization of development and genetic assimilation of acquired characters. *Nature*, 183, 1654–5.

23 CROGHAN,P.C. & LOCKWOOD,L.A.P.M. (1960) The composition of the haemolymph of the larva of *Drosophila melanogaster*. *J. Exptl. Biol.*, 37, 339–43.

24 BRIDGES,C.B. & BREHME,K.S. (1944) The mutants of *Drosophila melanogaster*. *Carnegie Inst. Wash. Publ. No.* 352.

25 WADDINGTON,C.H. (1940) The genetic control of wing development in *Drosophila*. *J. Genet.*, 39, 75–139.

26 WADDINGTON,C.H. (1940) *Organisers and Genes*. London and New York: Cambridge University Press.

27 WADDINGTON,C.H. (1941) Evolution of developmental systems. *Nature*, 147, 108–110, and paper 2 in this collection.

28 STERN,C. (1958) Selection for sub-threshold differences and the origin of pseudo-exogenous adaptations. *Am. Naturalist*, 102, 313–16.

29 STERN,C. (1959) Variation and hereditary transmission. *Proc. Am. Phil. Soc.*, 103, 183–9.

30 SCHMALHAUSEN,I.L. (1949) *Factors of Evolution*. New York: Blakiston (McGraw-Hill).

31 GOLDSCHMIDT,R.B. (1935) Gen und Ausseneigenschaft. *Z. induktive Abstammungs-u. Vererbungslehre*, 69, 38–131.

32 SANG,J.H. & McDONALD,J.M. (1954) The production of phenocopies in *Drosophila* using salts, particularly sodium metaborate. *J. Genet.*, 52, 392–412.

33 GOLDSCHMIDT,R.B. & PITERNICK,L.K. (1957) A genetic background of chemically induced phenocopies in *Drosophila*. *J. Exptl. Zool.*, 135, 127–202.

34 GOLDSCHMIDT,R.B. & PITERNICK,L.K. (1957) A genetic background of chemically induced phenocopies in *Drosophila* 2. *J. Exptl. Zool.*, 136, 201–28.

35 LANDAUER,W. (1957) Phenocopies and genotype, with special reference to sporadically-occurring developmental variants. *Am. Naturalist*, 91, 79–90.

36 LANDAUER,W. (1958) On phenocopies, their developmental physiology and genetic meaning. *Am. Naturalist*, 102, 201–13.

37 WADDINGTON,C.H. (1948) The concept of equilibrium in embryology. *Folia Biotheoret.*, B3, 127–38.

38 CANNON,W.B. (1932) *The Wisdom of the Body*. New York: Norton.

39 LERNER,I.M. (1954) *Genetic Homeostasis*. Edinburgh: Oliver and Boyd.

40 LEWONTIN,R.C. (1957) The adaptations of populations to varying environments. *Cold Spring Harbor Symposia Quant. Biol.*, 22, 395–408.

41 DOBZHANSKY,TH. (1955) A review of some fundamental concepts and problems of population genetics. *Cold Spring Harbor Symposia Quant. Biol.*,20,1–10.

42 THODAY,J.M. (1953) Components of fitness. *Symposia Soc. Exptl. Biol. No.* 7,96–113.

43 BEARDMORE,J.A., DOBZHANSKY,TH. & PAVLOVSKY,O.A. (1960) An attempt to compare the fitness of polymorphic and monomorphic experimental populations of *Drosophila pseudoobscura*. *Heredity*,14,19–33.

44 ROBERTSON,F.W. (1959) Gene-environment interaction in relation to the nutrition and growth of *Drosophila*. *Biol. Contribs. Univ. Texas, Publ. No.* 5914.

45 ROBERTSON,F.W. (1960) The ecological genetics of growth in *Drosophila*, I and II. *Genet. Research*,1,288–318.

46 FALCONER,D.S. (1960) Selection of mice for growth on high and low planes of nutrition. *Genet. Research.*1,91–113.

47 WADDINGTON,C.H. (1952) Canalization of the development of quantitative characters. *Quantitative Inheritance*,43–7. London: H.M.S.O.

48 WADDINGTON,C.H. (1955) On a case of quantitative variation on either side of the wild-type. *Z. induktive Abstammungs-u. Vererbungslehre*,87,208–28.

49 DUN,R.B. & FRASER,A.S. (1959) Selection for an invariant character, vibrissa number, in the house mouse. *Australian J. Biol. Sci.*,12,506–23.

50 FRASER,A.S., NAY,T. & KINDRED,B. (1959) Variation of vibrissa number in the mouse. *Australian J. Biol. Sci.*,12,331–9.

51 FRASER,A.S. & KINDRED,B. (1960) Selection for an invariant character vibrissa number, in the house mouse. II. Limits to variability. *Australian J. Biol. Sci.*,13,48–58.

52 RENDEL,J.M. (1959) Natural and artificial selection. *Australian J. Biol.Sci.*,22,22–7.

53 RENDEL,J.M. (1959) Canalization of the scute phenotype of *Drosophila*. *Evolution*,13,425–39.

54 RENDEL,J.M. & SHELDON,B.L. (1960) Selection for canalization of the scute phenotype in *Drosophila melanogaster*. *Australian J. Biol. Sci.*,13,36–47.

55 MAYNARD-SMITH,J. & SONDHI,K.C. (1960) The genetics of a pattern. *Genetics*,45,1039–50.

56 MAYNARD-SMITH,J. (1960) Continuous quantized and modal variation. *Proc. Roy. Soc. B.*,152,397–409.

57 WADDINGTON,C.H. (1960) Experiments on canalizing selection. *Genet. Research*,1,140–50.

58 MATHER,K. (1955) Genetical control of stability in development. *Heredity*,7,297–336.

59 JINKS,J.L. & MATHER,K. (1955) Stability of development of heterozygotes and homozygotes. *Proc. Roy. Soc. B.*,143,561–78.

60 TEBB,G. & THODAY,J.M. (1954) Stability in development and relational balance of X chromosomes in *Drosophila melanogaster*. *Nature*, 174, 1109–10.

61 THODAY,J.M. (1956) Balance, heterozygosity and developmental stability. *Cold Spring Harbor Symposia Quant. Biol.*, 21, 318–26.

62 THODAY,J.M. (1958) Homeostasis in selection experiment. *Heredity*, 12, 401–15.

63 REEVE,E.C.R. (1960) Some genetic tests on asymmetry of sternopleural chaetae number in *Drosophila*. *Genet. Research*, 1, 151–72.

64 REEVE,E.C.R. & ROBERTSON,F. (1954) Studies in quantitative inheritance. VI. Sternite chaeta number in *D. melanogaster*. *Z. induktive Abstammungs-u. Vererbungslehre*, 86, 269–88.

65 CLAYTON,G.A., ROBERTSON,A. & MORRIS,J.A. (1957) An experimental check on quantitative genetical theory. I. Short-term responses to selection. *J. Genet.*, 55, 131–51.

66 MATHER,K. (1943) Polygenic balance in the canalization of development. *Nature*, 151, 68–71.

67 WADDINGTON,C.H. (1943) Polygenes and oligogenes. *Nature*, 151, 394.

68 FRASER,A.S. (1960) Simulation of genetic systems by automatic digital computers: VI. Epistasis, and VII. Effects of reproduction rate and intensity of selection on genetic structure. *Australian J. Biol. Sci.*, 13, 150–62 and 344–50.

69 BERG,R.L. (1959) A general evolutionary principle underlying the origin of developmental homeostasis. *Am. Naturalist*, 93, 103–5.

70 SHATOURY,H.H. (1956) Developmental interactions in the differentiation of the imaginal muscles of *Drosophila*. *J. Embryol. Exptl. Morphol.*, 4, 228–39.

71 HUXLEY,J.S. (1942) *Evolution: The Modern Synthesis*. London: Al and Unwin.

72 SIMPSON,G.G. (1953) The Baldwin effect. *Evolution*, 7, 110–17.

73 MAYR,E. (1959) Where are we? *Cold Spring Harbor Symposia Quant. Biol.*, 24, 1–14.

74 WADDINGTON,C.H. (1953) The Baldwin effect, genetic assimilation and homeostasis. *Evolution*, 7, 386–7.

75 NAUMENKO,V.A. (1941) Fixation of certain mutations by artificial selection of corresponding modifications. *Compt. rend. acad. sci. U.R.S.S.*, 32, 1–15.

PAPER 11 (98–103)

1 WADDINGTON,C.H. (1940) *Organisers and Genes*. Cambridge: Cambridge University Press.

2 LANDAUER,W. (1948) *Genetics*, 33, 133.

3 LANDAUER,W. & BLISS,C.I. (1946) *J. exp. Zool.*, 102, 1.

PAPER 12 (103–15)

1 SCHMALHAUSEN,I.I. (1949) *Factors of Evolution*. Philadelphia: Blakiston.

2 WADDINGTON,C.H. (1940) *Organisers and Genes*. Cambridge: Cambridge University Press.

3 WADDINGTON,C.H. (1957) *The Strategy of the Genes*. London: Allen and Unwin.

4 MATHER,K. (1955) Genetical control of stability in development. *Heredity*, 7, 197.

5 TEBB, G. & THODAY,J.M. (1954) Stability in development and relational balance of X chromosomes in *D. melanogaster*. *Nature*, 174, 1109.

6 REEVE,E.C.R. & ROBERTSON,F.W. (1954) Studies in quantitative inheritance. VI. Sternite chaeta number in *Drosophila*; a metameric quantitative character. *Z. indukt. Abstamm. u-VererbLehre*, 86, 269.

7 FALCONER,D.S. (1957) Selection for phenotypic intermediates in *Drosophila*. *J. Genet.*, 55, 551.

8 FALCONER,D.S. & ROBERTSON,A. (1956) Selection for environmental variability of body size in mice. *Z. indukt. Abstamm. u-VererbLehre*, 87, 385.

9 STERN,C. & SHAEFFER,E.W. (1934) On the wild-type isoalleles in *D. melanogaster*. *Proc. nat. Acad. Sci.*, *Wash.*, 29, 361.

10 WADDINGTON,C.H., GRABER,H. & WOOLF,B. (1957) Isoalleles and the response to selection. *J. Genet.*, 55, 246.

11 WADDINGTON,C.H. & CLAYTON,R.M. (1952) A note on some alleles of aristopedia. *J. Genet.*, 51, 123.

PAPER 13 (116–26)

1 WADDINGTON,C.H. (1960) Experiments on canalizing selection. *Genet. Res.*, 1, 140-50.

2 THODAY,J.M. *et al.* (1959–64) Effects of disruptive selection, I–IX. *Heredity*, 13, 187–203, 205–18; 14, 35–49; 15, 119–217; 16, 219–23; 17, 1–27; 18, 513–24; 19, 125–30.

3 MAYNARD-SMITH,J. (1962) Disruptive selection, polymorphism and sympatric speciation. *Nature*, 195, 60–2.

4 THODAY,J.M. & GIBSON,J.B. (1962) Isolation by disruptive selection. *Nature*, 193, 1164–6.

5 THODAY,J.M. (1964) Genetics and the integration of reproductive systems. *Insect Reproduction* (Symp. No. 2, Roy. Entomol. Soc.), 108–20.

6 THODAY,J.M. (1964) Effects of selection for genetic diversity. *Genetics Today. Proc. 11th int. Congr. Genet.*, 533–40.

7 WALLACE,B. (1954) Genetic divergence of isolated populations of *D. melanogaster*. *Proc. 9th int. Congr. Genet.* (Caryologia Suppl.), 761–4.

8 KOOPMAN,K.F. (1950) Natural selection for reproductive isolation between *D. pseudoobscura* and *D. persimilis*. *Evolution*, 4, 135–48.

9 KNIGHT,G.R., ROBERTSON,A. & WADDINGTON,C.H. (1956) Selection for sexual isolation within a species. *Evolution*, 10, 14–22, and paper 15 in this collection.

10 SCHARLOO,W. (1964) The effect of disruptive and stabilizing selection on the expression of a cubitus interruptus mutant in *Drosophila. Genetics,* 50, 553–62.

PAPER 14 (128–37)

1 THORPE,W.H. (1945) *J. Anim. Ecol.,* 14, 67.
2 WADDINGTON,C.H. (1953) *Symp. Soc. Exp. Biol.,* 7, 186.
3 JANZER,W. & LUDWIG,W. (1953) *Zeits. und Abst. Vererb.,* 84, 462.
4 HOVANITZ,W. (1948) *Contrib. Lab. Vert. Biol. Michigan, No. 41.*
5 HOVANITZ,W. (1953) *Symp. Soc. Exp. Biol.,* 7, 238.
6 SIMPSON,G.G. (1953) *Evolution,* 7, 110.
7 WADDINGTON,C.H. (1953) *Evolution,* 7, 118.

PAPER 15 (138–50)

1 MULLER,H.J. (1939) Reversibility in evolution considered from the standpoint of genetics. *Biol. Rev.,* 14, 261–80.
2 DOBZHANSKY,T. (1937) *Genetics and the Origin of Species.* New York: Columbia University Press.
3 MERRELL,D.J. (1954) Sexual isolation between *Drosophila persimilis* and *Drosophila pseudoobscura. Am. Nat.,* 88, 93–100.
4 KOOPMAN,K.F. (1950) Natural selection for reproductive isolation between *Drosophila pseudoobscura* and *Drosophila persimilis. Evolution,* 4, 135–48.
5 WALLACE,B. (1950) An experiment on sexual isolation. *D.I.S.,* 24, 94–6.
6 RENDEL,J.M. (1951) Mating of ebony, vestigial and wild-type *Drosophila melanogaster* in light and dark. *Evolution,* 5, 226–30.
7 WALLACE,B. (1954) *Proc. 9th int. Congr. Genet.* (Caryologia Suppl.), 761–4.
8 DOBZHANSKY,T. & KOLLER,P.C. (1938) An experimental study of sexual isolation in *Drosophila. Biol. Zentral.,* 58, 589–607.
9 KING,J.C. (1947) Interspecific relationships within the guarani group of *Drosophila. Evolution,* 1, 143–53.
10 WADDINGTON,C.H., WOOLF,B. & PERRY,M. (1954) Environment selection by *Drosophila* mutants. *Evolution,* 8, 89–96, and paper 14 in this collection.

PAPER 16 (150–61)

1 MULLER,H.J. (1949) The Darwinian and modern concept of natural selection. *Proc. Amer. Phil. Soc.,* 93, 459–70.
2 DOBZHANSKY,TH. (1941) *Genetics and the Origin of Species.* New York: Columbia University Press.
3 MERRELL,D.J. (1949) Selection mating in *Drosophila melanogaster. Genetics,* 34, 370–89.
4 DOBZHANSKY,TH. & MAYR,E. (1944) Experiments on sexual isolation. I. Geographic strains of *D. willistoni. Proc. Nat. Acad. Sci.,* 30, 238–44.

5 DOBZHANSKY,TH. (1946) Experiments on sexual isolation in
   Drosophila. VII. The nature of isolating mechanisms between *D.
   pseudoobscura* and *D. persimilis*. *Proc. Nat. Acad. Sci.*,32,128–37.
6 KOOPMAN,K.F. (1950) Natural selection for reproductive isolation
   between *Drosophila pseudoobscura* and *Drosophila persimilis*. *Evolution*,
   4,135–48.
7 KNIGHT,G.R., ROBERTSON,A. & WADDINGTON,C.H. (1956)
   Selection for sexual isolation within a species. *Evolution*,10,14–22,
   and paper 15 in this collection.
8 MERRELL,D.J. (1949) Mating between two strains of *Drosophila
   melanogaster*. *Evolution*,3,266–8.
9 RENDEL,J.M. (1944) Genetics and cytology of *Drosophila
   subobscura*. II. Normal and selective matings in *D. subobscura*.
   J. Genet.,46,287–302.
10 BASTOCK, MARGARET (1956) A gene mutation which changes a
   behaviour pattern. *Evolution*,10,421–39.
11 MAYR,E. & DOBZHANSKY,TH. (1945) Experiments on sexual
   isolation in *Drosophila*. IV. Modification of the degree of isolation
   between *Drosophila pseudoobscura* and *D. persimilis* and of sexual
   preferences in *D. prosaltans*. *Proc. Nat. Acad. Sci.*, 31, 75–82.

PAPER 20 (183–92)
1 PATTEE,H.H. (1966) Physical theories, automata, and the origins
   of life. *Natural Automata and Useful Simulations*, 76. Washington:
   Spartan Books and London: Macmillan.
2 *Mathematical Challenges to the Neo-Darwinian Interpretation of Evolution.*
   Wistar Institute Symposium monograph no. 5, 1967.
3 Ibid., p. 74.
4 ZEEMAN,E.C. & BUNEMAN,O.P. (1968) Tolerance spaces and
   the brain. *Towards a Theoretical Biology* 1: *Prolegomena* (ed.
   Waddington,C.H.),140–51. Edinburgh: Edinburgh University
   Press.
5 BOSSERT,W. (1967) Mathematical optimization: Are there abstract
   limits on natural selection? *Mathematical Challenges to the
   Neo-Darwinian Interpretation of Evolution*. Wistar Institute Symposium
   monograph no. 5.
6 Ibid., p. 90.
7 Ibid., p. 11.

PAPER 21 (192–7)
1 WADDINGTON,C.H. (1957) *The Strategy of the Genes*. London: Allen
   and Unwin.
2 LEWONTIN,R.C. (1961) Evolution and the theory of games.
   J. Theoret. Biol.,1,382.
3 KACSER,H. (1963) The kinetic structure of organisms. *Biological
   Organization at the Cellular and Supercellular Level* (ed. Harris,R.J.C.),
   25–41. New York: Academic Press.

PAPER 23 (201–9)

1 CAIRNS SMITH,A.G. (1965) The origin of life and the nature of the primitive gene. *J. Theor. Biol.*, **10,** 53–88.

2 HALDANE,J.B.S. (1924) The mathematical theory of natural and artificial selection. Part I, *Cambridge Philos. Soc.*, **23,** 19–41.

PAPER 24 (209–30)

1 Actually, it turns out that some degree of order—not real buffering, but perhaps akin to it—does emerge if each component interacts with a *small* number of other components. See S.Kauffman in *Towards a Theoretical Biology 4: Essays,* 243 ff. (Edinburgh University Press 1972).

PAPER 25 (231–52)

1 Wheldon's polemical journal, *Questions of the Day and of the Fray,* did not cease publication until 1924 (?); and the anti-Mendelian influence was strong enough to ensure that Fisher's first major genetical paper, arguing amongst other things that continuous variation is explicable on Mendelian principles, was rejected by the Royal Society of London, and published by the less prestigeful Royal Society of Edinburgh in 1918.

2 MAYNARD-SMITH,J. (1969) The status of neo-Darwinism. *Towards a Theoretical Biology 2: Sketches* (ed. Waddington,C.H.),82–9. Edinburgh: Edinburgh University Press.

3 MAYNARD-SMITH,J. (1967) The counting problem. *Towards a Theoretical Biology 1: Prolegomena* (ed. Waddington,C.H.),120–4. Edinburgh: Edinburgh University Press.

4 WRIGHT,S. (1964) Stochastic processes in evolution. *Stochastic Models in Biology and Medicine* (ed. Garland),199. University of Wisconsin Press.

5 LEVINS,R. (1962) *Amer. Nat.*, **96,** 361; (1963) *Amer. Nat.* **97,** 75; (1964) *J. Theoret. Biol.*, **7,** 224; (1965) *Evolution,* **18,** 635; (1965) *Genetics,* **52,** 891.

6 MACARTHUR,H.H. (1962) *Proc. Nat. Acad. Sci. Wash.*, **48,** 1893.

7 SLOBOTKIN,L.B. (1964) *Am. Sci.*, **52,** 343.

8 WADDINGTON,C.H. (1961) *The Nature of Life,* 109. Allen and Unwin. Also (1967) in *Towards a Theoretical Biology 1: Prolegomena* (ed. Waddington,C.H.),22. Edinburgh: Edinburgh University Press.

9 Several ecologists argue that there will be, for ecological reasons, a selection pressure towards the development of more complex ecosystems, involving an ever-increasing number of species. The fundamental selection is towards increasing the stability of the ecosystem; when there are only a small number of interacting species, there is a tendency for violent fluctuations in numbers, which may lead to the extinction of some species and the complete collapse of the whole system. See, for example, Hutchinson,G.E.

(1959) *Amer. Nat.*, **93**, 145; MacArthur, R. H. (1955) *Ecology*, **36**, 533; Dunbar, M. J. (1968) *Ecological Development in Polar Regions*, Prentice Hall.

10 GREEN, M. M. (1965) *Proc.* 11*th int. Cong. Genet.*, **2**, 37.
11 For this and other examples see *The Genetics of Colonizing Species*. IUBS Symposium (ed. Baker and Stebbins). Academic Press, 1965.
12 PATTEE, H. H. (1967) The physical basis of coding and reliability in biological evolution. *Towards a Thoeretical Biology* 1 : *Prolegomena* (ed. Waddington, C. H.), 67–93. Edinburgh: Edinburgh University Press.

13 In connection with the nature of the mutational events which may play a part in evolution I would like to advance some ideas, which are, I freely admit, so speculative that I have relegated them to a note rather than the body of the text; but many biologists will consider them so outrageously heterodox that they may refuse to consider them at all—and these I should beg to think again.

One of the major problems of evolution theory is to understand how the sharp discontinuities between major taxonomic groups— Phyla, Families, Species-Groups, and so on—have come into being. A simple-minded empirical inspection of the facts would suggest, as it did for instance to Goldschmidt when he wrote his *The Material Basis of Evolution* 1940 and *Theoretical Genetics* 1955, that it might be profitable to contemplate the possibility of the very occasional occurrence of what Goldschmidt called 'systemic mutations', which result in a complete restructuring of the genome, achieved either in a single step, or at least in rather few generations. When Goldschmidt wrote, no clear-cut example could be given in which the occurrence of such a process had been observed. The orderly minded orthodox biological world closed its ranks against this suggestion that revolutionary processes may happen. It became accepted that the only respectable doctrine is that evolution never involves anything but step-by-step Fabian gradualism, plodding along a weary way similar to that by which the annual milk yield of dairy cows or egg yield of hens is slowly improved—the occurrence of a little allopolyploidy or rearrangement of chromosomes by two or three breaks could be admitted, but would only push the basic philosophy from Bourgeois-Liberal to right-wing Social Democrat.

It is still impossible—so far as I know—to quote a compelling instance in which a systemic mutational event has been observed in an evolving multicellular organism. But events which appear to be essentially of this kind are becoming well known in the field of cell culture. It is a common experience that cells isolated from vertebrate tissues usually grow in culture for a fairly restricted number of cell generations—a hundred or two—and then die out, *unless* they undergo some sort of change which brings into being cells capable of forming an 'established line', which can then be sub-cultured in perpetuity. The nature of the change from a 'strain' to an 'established line' is highly obscure, but it often involves what looks like a complete restructuring of the genome; there may be a considerable reduction in number of chromosomes, accompanied in some cases by considerable changes in chromosome morphology. (For a recent review see *The mammalian cell as a differentiated microorganism* by Howard Green and George J. Todaro (1967) *Ann. Rev. Microbiol.*, 21, 574–600.)

The fact that cells in culture can throw up, within at most a few cell generations, new types of cells capable of giving rise to 'established lines', and that the change may involve a very drastic reshuffling of the genome (usually, in the cases observed, with a loss rather than a gain of chromosomal material), is evidence that something like a 'genetic revolution' or 'systemic mutation' can occur. It is, of course, more difficult to see how such an event in an evolving population could be propagated so as to affect the future, but if such events are not ruled out of court by the nature of genetic processes it seems silly to close one's mind to the possibility that evolution has found some way of making use of them.

An example of an evolutionary phenomenon which suggests a very radical reorganization of the genome between nearly related species is the astonishing difference found by Forbes Robertson in the DNA sequences of *Drosophila melanogaster* and *simulans*, as tested by molecular hybridization. RNA manufactured *in vitro* on a *melanogaster* DNA template hybridizes only one-third as well with *simulans* DNA as does the RNA made on the *simulans* template; the results of a reciprocal experiment are very similar. Although it seems certain that the hybridizations only occur between substances related to the highly reiterated stretches of DNA, this is still strongly suggestive that the differences between these two species involve much more radical and pervasive alterations of base sequences than were contemplated a few years ago, when it seemed that all that was involved was a few inversions, and translocations of large sections of the chromosomes.

14 LEWONTIN, R. C. (1966) *Bioscience*, 16, 25.
15 BARNET, S. A. (1965) *Biol. Rev.*, 40, 5.
PAPER 26 (253–66)
1 THOM, R. (1970) Topological models in biology. *Towards a Theoretical Biology 3: Drafts* (ed. Waddington, C. H.), 88–116. Edinburgh: Edinburgh University Press.
2 ZEEMAN, E. C. (1972) Differential equations for the heartbeat and nerve impulse. *Towards a Theoretical Biology 4: Essays* (ed. Waddington, C. H.), 8–67. Edinburgh: Edinburgh University Press.
3 FISHER, R. A. (1930) *The Genetical Theory of Natural Selection*. Oxford: Clarendon Press.

4 WADDINGTON,C.H. (1957) *The Strategy of the Genes.* London: Allen and Unwin.
5 WADDINGTON,C.H. (1942) Canalization of development and the inheritance of acquired characters. *Nature*, 150, 563, and paper 3 in this collection,
6 WADDINGTON,C.H. (1953) Genetic assimilation of an acquired character. *Evolution*, 7, 386–7.
7 WADDINGTON,C.H. (1961) Genetic assimilation. *Ad. Genet.*, 12, 257–93, and paper 8 in this collection.

PAPER 27 (266–81)

1 NEEDHAM,J. (1936) *Order and Life.* Yale University Press.
2 WADDINGTON,C.H. (1956) *Principles of Embryology.* London: Allen and Unwin.
3 The expression of this idea in English poetry is discussed by E. M. W. Tillyard (1943) *The Elizabethan World Picture.* London: Chatto & Windus.
4 RENSCH,B. (1959) *Evolution above the Species Level.* London: Methuen. Translation and partial revision of *Neuere Probleme der Abstammungslehre*, 2nd edition. Stuttgart: Enke Verlag, 1947. For end-directed and teleonomic (adaptive) evolution, see C. S. Pittendrigh (1958) Adaptation, selection and behaviour, in *Behaviour and Evolution* (eds. Roe, A. & Simpson, G. G.). Yale University Press.
5 'Psychosocial' is Huxley's word. To my mind, it suffers from some redundancy, since the social can hardly avoid being psychological. I prefer to use 'sociogenetic', which emphasizes the importance of the mechanism as a means of transmitting information from one generation to the next, which is the crucial point.
6 WADDINGTON,C.H. Evolutionary systems—animal and human. *Nature*, 183, 1634–8; Huxley, J. S. (1955) Evolution, cultural and biological, *Yearbk. of Anthropol.* (ed. Morris); Rensch, B. (1959) Homo Sapiens, *Vom Tier zum Halbgott*, Gottingen.
7 MULLER,H.J. (1960) The guidance of human evolution. *Evolution After Darwin*, vol. 2 *Evolution of Man*, Chicago University Press; Teilhard de Chardin, P. (1959) *The Phenomenon of Man*, London: Collins.
8 DARLINGTON,C.D. (1939) *The Evolution of Genetic Systems.* Cambridge University Press.
9 PONTECORVO,G. (1959) *Trends in Genetic Analysis.* Columbia University Press.
10 See *Animal Behaviour* (ed. Thorpe, W. H.). Cambridge University Press.
11 These arguments are more fully developed in a recent book, *The Ethical Animal.*
12 Discussed in Langer, S. (1942) *Philosophy in a New Key*, chapter 3. Harvard University Press.

13 This postscript was published in the journal *Theoria to Theory*, 2
    (April 1968) 240–1.
14 HARDY, Sir A. (1965) *The Living Stream*. London: Collins.
15 The numbered points are those made by Lloyd Morgan and quoted
    by Sir Alister Hardy on page 167 of *The Living Stream*. They read:
    8. Let us suppose, however, that a group of organisms belonging to a
    plastic species is placed under new conditions of environment.
    9. Those whose innate somatic plasticity is equal to the occasion
    survive. They are modified. Those whose innate plasticity is not equal
    to the occasion are eliminated.
    10. Such modification takes place generation after generation but, as
    such, is not inherited. There is no transmission of the effects of
    modification to the germinal substance.
    11. But variations in the same direction as the somatic modification
    are now no longer repressed and are allowed full scope.
    12. Any congenital variations antagonistic in direction to these
    modifications will tend to thwart them and render the organism in
    which they occur liable to elimination.
    13. Any congenital variations similar in direction to these modifications
    will tend to support them and to favour the individuals in which they
    occur.
    14. Thus will arise a congenital predisposition to the modifications in
    question.
16 See paper 14 in this collection.
17 See paper 17 in this collection.

PAPER 28 (281–99)
 1 HOFSTADTER, R. (1955) *Social Darwinism in American Thought*.
    Revised ed. Boston: Beacon Press.
 2 HORAWITZ, W. (1952) *Symp. Soc. Exp. Biol.*, 7, 238.
 3 WADDINGTON, C.H., WOOLF, B. & PERRY, M.M. (1954)
    Evolution, 8, 89, and paper 14 in this collection.
 4 KOOPMAN, K.F. (1950) *Evolution*, 4, 135.
 5 KNIGHT, G.R., ROBERTSON, A. & WADDINGTON, C.H. (1956)
    Evolution, 10, 14, and paper 15 in this collection.
 6 WADDINGTON, C.H. (1957) *The Strategy of the Genes*. London: Allen
    and Unwin.
 7 DARLINGTON, C.D. (1946) *The Evolution of Genetic Systems*, 2nd
    edition. Cambridge University Press.
 8 MEAD, M. (1958) Cultural determinants of behaviour. *Behaviour and
    Evolution* (eds. Roe, A. & Simpson, G.G.). Yale University Press.

PAPER 29 (299–307)
 1 HALDANE, J.B.S. (1932) *The Causes of Evolution*. London:
    Longmans Green.
 2 HALDANE, J.B.S. (1955) Population genetics. *New Biol.*, 18, 51.

3 MAYR,E. (1970) *Populations, Species and Evolution*. Harvard University Press.
4 HAMILTON,W.D. (1964) The genetical evolution of social behaviour. *J. Theoret. Biol.*, 7, 1–16, 17–52.
5 MAYNARD-SMITH,J. (1964) Group selection and kin selection. *Nature*, 201, 1145–7.
6 MAYNARD-SMITH,J. (1972) *On Evolution*. Edinburgh: Edinburgh University Press.
7 WILLIAMS,G.C. (1966) *Adaptation and Natural Selection*. Princeton University Press.
8 MAYNARD-SMITH,J. & PRICE,G.E. (1973) The logic of animal conflict. *Nature*, 246, 15–18.
9 HINDE,R.A., ed. (1972) *Non-Verbal Communication*. Cambridge University Press.
10 LANGER, SUZANNE (1967, 1972) *Mind, a Study of Human Feeling*. Baltimore: Johns Hopkins Press.
11 CHOMSKY,N. (1965) *Aspects of the Theory of Syntax*. M.I.T. Press.
12 WADDINGTON,C.H. (1974) The evolution of mind. *The Development of Mind*. Edinburgh: Edinburgh University Press.

ACKNOWLEDGEMENTS

1 The practical consequences of metaphysical beliefs on a biologist's work. An autobiographical note. *Towards a Theoretical Biology 2: Sketches* (ed. C.H.Waddington) 72–81. Edinburgh University Press 1969.

2 The evolution of developmental systems. *Nature* **147** (1941) 108.

3 Canalization of development and the inheritance of acquired characters. *Nature* **150** (1941) 563.

4 Selection of the genetic basis for an acquired character. *Nature* **169** (1952) 278.

5 The evolution of adaptations. *Endeavour* xii no.47 (1953).

6 Evolution and epistemology. *Nature* **173** (1954) 880.

7 Evolutionary adaptation. *Evolution after Darwin*, University of Chicago Centennial, 381–402. University of Chicago Press 1959.

8 Genetic assimilation. *Advances in Genetics* **10** (1961) 257–90.

11 Canalization of the development of quantitative characters. *Quantitative Inheritance* (eds. E.C.R.Reeve & C. H.Waddington) 43–6. H.M.S.O. 1952.

12 Experiments on canalizing selection. *Genetic Research* **1** (1960) 140–50.

13 With E.Robertson: Selection for developmental canalization. *Genetic Research* **7** (1966) 303–12.

14 With B.Woolf & M.M.Perry: Environment selection by Drosophila mutants. *Evolution* viii no.2 (1954) 89–96.

15 With G.R.Knight & A.Robertson: Selection for sexual isolation within a species. *Evolution* x no.1 (1956) 14–22.

16 With S.Koref Santibañez: The origin of sexual isolation between different lines within a species. *Evolution* xii no.4 (1958) 485–93.

18 a) Behaviour as a product of evolution. *Science* **129** (1959) 203.
   b) Large scale evolution. *Discovery*, October 1960.
   d) The resistance to evolutionary change. *Nature* **175** (1955) 51.
   e) Individual paradigms and population paradigms. *TLS*, 22 October 1971.

19 With R.C.Lewontin: A note on evolution and changes in the quantity of genetic information. *Towards a Theoretical Biology 1: Prolegomena* (ed. C.H.Waddington) 109–10. Edinburgh University Press 1968.

20 Does evolution depend on random search? *Towards a Theoretical Biology 1: Prolegomena* (ed. C. H. Waddington) 111–19. Edinburgh University Press 1968.

21 Colonizing species. *The Genetics of Colonizing Species*. Academic Press 1965.

22 The principle of archetypes in evolution. *Mathematical Challenges to the Neo-Darwinian Interpretation of Evolution*. The Wistar Symposium Monograph No. 5, 1967.

23 The evolutionary process. *Population Biology and Evolution*, 37–45. Syracuse University Press 1968.

24 The basic ideas of biology. *Towards a Theoretical Biology 1: Prolegomena* (ed. C. H. Waddington) 1–32. Edinburgh University Press 1968.

25 Paradigm for an evolutionary process. *Towards a Theoretical Biology 2: Sketches* (ed. C. H. Waddington) 106–28. Edinburgh University Press 1969.

26 A catastrophe theory of evolution. *Annals of the New York Academy of Science 1974* (in press).

27 The human animal. *The Humanist Frame* (ed. J. Huxley). George Allen and Unwin 1961.

28 The human evolutionary system. *Darwinism and the Study of Society* (ed. M. Banton) 63–81. Tavistock Publications 1961.

Index